I0490481

SP🎾RTS
&
MATHEMAT🏀CS

SP🎾RTS
&
MATHEMAT🏀CS

Leisure 🏐 Learning

We are living in an era that athletic talent
and math intellect are valued and rewarded

Reza Noubary

2021

Copyright © 2021 by Reza Noubary.

Library of Congress Control Number: 2020923339
ISBN: Hardcover   978-1-6641-4348-7
                   Softcover   978-1-6641-4349-4
                   eBook   978-1-6641-4350-0

All rights reserved. No part of this book may be reproduced or transmitted in any form or by any means, electronic or mechanical, including photocopying, recording, or by any information storage and retrieval system, without permission in writing from the copyright owner.

Any people depicted in stock imagery provided by Getty Images are models, and such images are being used for illustrative purposes only. Certain stock imagery © Getty Images.

Print information available on the last page.

Rev. date: 12/03/2020

To order additional copies of this book, contact:
Xlibris
844-714-8691
www.Xlibris.com
Orders@Xlibris.com
821933

Contents

DEDICATION

I dedicate this book to my beautiful
granddaughter SHIELA with the
hope that she will develop
appreciation for this
type of effort.

ABOUT COVER PICTURE

THE CURVE IN the cover picture is the well-known Golden Spiral, which is related to golden ratio and Fibonacci numbers. Chances are, you've heard of the Golden Ratio in school, in mathematics, art, or design class. Perhaps you saw the movie _The DaVinci Code_ and learned about it.

One of the reasons the Golden Spiral is such an effective compositional tool is its prevalence in nature. Organic examples of the Golden Spirals can be found throughout the natural world. Nautilus shells, sunflowers, and pinecones are a handful of readily recognizable examples. Many people find images composed utilizing the Golden Spiral to have an organic and aesthetically pleasing quality to them.

ACKNOWLEDGMENT

My gratitude to my wife extraordinaire, Zohreh, who always willingly helps me to do my work. I cannot fail to honor the memory of my mother and brother.

I would like to acknowledge my colleagues Bill Calhoun and Drue Coles for the co-authorship of the articles "Changing the Rules of Tennis: An Exercise in Mathematical Modeling" and "Rule of Tangent for Win-By-Two Games," respectively.

Special thanks to Steven Cohen, Dong Zhang, and Youmin Lu for their constant assistant and encouragement, and to Sergio Lemus for his help whenever I needed it.

ABOUT THE AUTHOR

THE AUTHOR WAS born to an Azari family in early 1946. He was the youngest of a clan of eleven children, some of whom died in early ages because of the lack of access to medical care. His father was a police officer who struggled with drug addiction and alcoholism. His mother had no formal education as she was forced to marry at a very young age.

Although not easy, he managed to go through formal education and receive his BSc and MSc in Mathematics from Tehran University, and MSc and PhD from Manchester University in England. He worked in several different universities in several different countries. He has also been a visiting scholar at Harvard, Princeton, U-Penn, UCLA, University of Maryland, University of Kaiserslautern, and Catholic University of Leuven. His research interests include risk analysis of natural disasters and applications of mathematics and statistics in sports. He is a fellow of the Alexander von Humboldt Foundation and a fellow or member of numerous professional organizations. He has published several scientific books and over one hundred research articles in more than ten different disciplines. His outside interests include music, soccer, racquetball, and tennis.

Socially he has experienced life as an insider, outsider, majority, minority, winner, loser, believer, denier, single, married, student, teacher, son, father, grandfather, uncle, friend, and enemy. He has dealt with a hard childhood, poverty, health issues both physical and mental, revolution, war, shortage, and stress of learning new languages and adjusting to different cultures. He has two sons and a granddaughter and lives with his wife in a peaceful small town in rural Pennsylvania.

PREFACE

SPORTS PROVIDE AN inexhaustible source of fascinating and challenging scientific problems. Today, most sports can be studied from a mathematical perspective to yield more valid quantitative results. For example, mathematical methods are applied to estimate an athlete's chances of success, identify the best training conditions, and measure their effectiveness. Game theory is used to develop strategies for players and coaches. Information theory makes it possible to estimate the amount of eyestrain in mountain skiing, table tennis, etc. Mathematical physics is used to identify the best shape of rowboats and oars. Applied probability and statistics has been instrumental in analyses of vast amounts of sport data for decision making. Probabilistic Monte Carlo methods are used for modeling and simulation of popular sports.

The athletic competitions also provide mathematicians with a wealth of research material. There is plenty of opportunity to experiment, test mathematical models and optimal strategies for situations occurring in sports. Only a tiny part–quite possibly not the most intriguing one–of the problems arising in sports has been described in the pages of mathematics books and journals. Think how many yet unsolved problems arise in different sports. Because of this, over the past few decades, the distinctive characteristics of traditional and high-level sports competitions have attracted the interest of the scientific community. Some of these studies have been directed at the modeling and analysis of the characteristics of different sports and some at the analysis of extraordinary performances and records.

Other than research, the universal popularity of sports has inspired a gold mine of interesting examples for teaching and learning. It is generally recognized that the use of sports marks an exciting new direction in teaching and learning mathematics and related subjects. With the present state of education, publications that connect mathematics to popular activities like sports are much needed. In fact, this has also been recognized and emphasized by the three major professional organizations: the American Mathematical Society, the Mathematical Association of America, and the American Statistical Association.

The goal of this book is to find a way to delight sport lovers about mathematics and mathematicians about sports to help them see their connections. It is hoped to bring a variety of applications within the reach of individuals with some mathematics background or interests. The book is appealing to teach from as well as to learn from as youth today show interest and enthusiasm for sports.

Teaching and Learning Values of Sports

- Sports have a general appeal, and it is an area to which modern scientific methods are increasingly applicable.
- Sports have become a part of everyday life, especially for young people.
- Young people usually enjoy sports and show a great deal of interest in mathematics and statistics applied to them.
- A major part of calculus and statistics sequences offered in schools and colleges can be taught using a chosen sport.

- Most young learners can relate to sports and can understand the rules and meanings of the different statistics presented to them.
- Sports data offer a unique opportunity to test methodologies offered by mathematics and statistics.
- I believe it is hard to find an area other than sports where one could collect reliable data with the highest precision possible.
- Almost all other data-producing disciplines are susceptible to "data mining" and error, since, unlike sports, they are not watched by millions of fans and media.

CHAPTER 1

Introduction

THIS INTRODUCTORY CHAPTER includes definitions of sport and mathematics, their relation and its teaching and research values.

What Is Sport?

Sport is often defined as an activity involving physical exertion and skill in which an individual or team competes against another or others for a title or entertainment. According to the Council of Europe Charter on Sport, it includes/means all forms of physical activities, which, through casual or organized participation, aim at expressing or improving physical fitness and mental well-being, forming social relationships, or obtaining results in competition at all levels. In other words, it refers to activities that, at least in part, aim to use, maintain, or improve physical ability and skills while providing enjoyment to participants, and in some cases, entertainment for spectators.

Classification of Sports

As can be anticipated, a large number of sports are played around the world. The most well-known sports may be classified into three groups: combat sports, object sports, and independent sports. A combat sport is one in which each competitor tries to control the opponent by direct confrontation (e.g., boxing, wrestling, and fencing). An object sport is one in which each competitor tries to control an object, while the other competitor is in direct confrontation (e.g., soccer, baseball, and chess). An independent sport is one in which one competitor may not interfere with the other competitor (e.g., swimming, shooting, and golf). In an independent sport, the competitors may perform at different times and even in different places, and it might be said that each competitor tries to control himself/herself.

There are three ways in which performance is evaluated in sports: judged (e.g., all combat sports, diving, and gymnastics), measured objectively (e.g., weightlifting, and swimming), or scored objectively (e.g., baseball, archery, and golf).

Sport in Schools

It has been argued that schools in the United States spend too much time and money on sports, much more than on academics. Some argue that if what schools do here was good for students, other countries, especially European countries, would have adopted it long ago. Although these arguments have some merit and relevance, there is also a positive side to sports in school. My son played soccer in school, and while in school, he took courses in college where I taught and earned fifty-six credits before graduating a valedictorian. He then went to graduate school and finally received his PhD from Harvard. When I asked him about high school experience and the part he enjoyed best, he said without a doubt playing soccer. He talked about the good feeling of being a part of a team, winning, losing, and sharing the happiness or sadness. He valued highly the feeling of being someone, representing a school, and making friends for life. What I like to add to this relates to students who are not academically strong. My son always felt good about himself simply because he was the best in the classroom. Participating in sport provides opportunity for students to experience being good or even the best. In addition, it is an avenue for students to release extra energy, become calm, and feel good about themselves.

What Is Mathematics?

Mathematics is a science with the ability to abstract and generalize. It is a unique tool for getting insight into any disciplines to which it is being applied. The story of mathematics is fascinating. Its history and philosophy provide an invaluable perspective on human nature and the world around them. It can be studied in its own right (pure mathematics) or as is applied to other disciplines (applied mathematics). It is known among mathematicians that "the relations between pure and applied mathematicians are based on trust and understanding." "Pure mathematicians do not trust applied mathematicians, and applied mathematicians do not understand pure mathematicians." Mathematicians often seek out patterns and use them to formulate new conjectures. They then resolve the truth or falsity of them by mathematical proof. When outcomes seem to be good models of real phenomena, mathematical reasoning is used to provide insight about the situation. A mathematical model is an abstract model that uses symbolic language to describe the behavior of a system. It is a representation of the essential aspects of a system in usable form. They are used for simulating real-life situations to forecast their future behavior. Of course, models are not the same as the real thing.

In "What is Mathematics? The Most Misunderstood Subject" by Dr. Robert H. Lewis, it is said that for more than two thousand years, mathematics has been a part of the search for understanding the world. Mathematical discoveries have come both from the attempt to describe the natural world and from the desire to arrive at a form of inescapable truth from careful reasoning. Meaningful mathematics is like journalism–it tells an interesting story. But unlike some journalism, the story has to be true. It is the study of quantity, structure, space, and change. The truth is established by rigorous deduction from appropriately chosen axioms and definitions.

References

Lewis, Robert H. "What is Mathematics? The Most Misunderstood Subject." https://www.fordham.edu/info/20603/what_is_mathematics.

"Persecution and the Art of Writing Pt V–The Literary Form of the Kuzari." http://search-for-emes.blogspot.com/2007_11_01_archive.html.

Let me explain

Most people relate to mathematics through numbers and often refer to their manipulation as mathematics and the manipulators as mathematician. Numbers manipulation, in fact, has not much to do with mathematics. If so, my calculator would be a great mathematician. Mathematics is partly a symbolic language for expressing and communicating complex ideas and relationships, and partly a perfect world arrived at by smoothing the rough world. Most of the smooth curves in calculus book only represent the ideal world or our expectations but not the real world.

Recall that ordinary language is incapable of describing, expressing, or explaining complex scientific ideas and concepts. Additionally, no language is universal. Since mathematics is a manmade science, it is unique and free of the uncertainties of the real world. In fact, it is the only discipline where theorems and proofs and deductive reasoning is used with zero margin of error. Other disciplines are mostly based on observation or experimentations where generalization takes place using inductive reasoning. As such, there is always a margin of error. Physicist and engineers use mathematics to model specific problem much like a tailor who makes suit. Mathematicians are like tailors who make all types of suits and let the user pick the one that fits the situation the best.

Mathematics is a concise language with well-defined rules for manipulations. Mathematical model is a description of a system using this language. The modeling includes methods of simulating real-life situations with mathematical equations to forecast their future behavior. This definition suggests that modeling is a cognitive activity in which we think about and make models to describe devices, objects, entities, or state of affairs. It is a representation of the essential aspects of systems to represent knowledge of that system in usable form.

References

https://www.slideshare.net/josealbertonohnoh/an-introduction-to-stochastic-modeling?cv=1.

Mathematics as a Language

Let me start with a quote from Galileo Galilei, "The Book of Nature is written in the language of mathematics." According to *Literacy Strategies for Improving Mathematics Instruction* by Joan M. Kenney, Euthecia Hancewicz, Loretta Heuer, Diana Metsisto and Cynthia L. Tuttle, there are over four thousand languages and dialects in the world, and all of them have one thing in common: they are instruments for communicating based on sounds or conventional symbols, words, and sentences. Most also have a category for words representing nouns, or objects, and a category for words representing verbs, or actions. The more-developed languages are also described in terms of a vocabulary of symbols or words, a grammar consisting of rules of how they may be used, and a syntax or propositional structure, which places the symbols in linear structures.

Literacy Strategies for Improving Mathematics Instruction also says that taking the commonality of the major languages as a starting point provides an interesting way of looking at the mathematical world and its language. One model proposed in 1995 suggests that we think about mathematical nouns, or objects, as being numbers, measurements, shapes, spaces, functions, patterns, data, and arrangements—items that comfortably map onto commonly accepted mathematics content strands. Mathematical verbs may be regarded as the four predominant actions that we ascribe to problem-solving and reasoning: modeling and formulating, transforming and manipulating, inferring, and communicating. Taken as a whole, these four actions represent the process that we go through to solve a problem.

As we know, a part of the English language is used for making formal mathematical statements and communicating definitions, theorems, proofs, word problems, and examples. Although English language is a source of knowledge, it is not designed for doing mathematics. Mathematics is usually written in a symbolic language that is designed to express mathematical ideas and thoughts.

The symbolic language used to present and communicate mathematics is a special-purpose language. "The Symbolic Language of Math" by Charles Wells states it has its own symbols and rules of grammar that are quite different from those of English. The language consists of symbolic expressions written in the way mathematicians traditionally write them. A symbol is a typographical character. The symbolic language also includes symbols that are specific to mathematics. We can usually read expressions in this language in any mathematical article written in any language.

Here are some elements of symbolic language from the article "The Language of Mathematics": 10 digits: $0, 1, 2, \ldots 9$. Symbols for operations: $+$, $-$, x, $/$. Symbols that "stand in" for values: x, y, z, . . . Special symbols like ϖ, $=$, $<$, $\leq$, . . . Here, nouns could be fixed things, such as numbers or expressions with numbers: 73, 5 $(3 - 1/7)$. The verb could be the equals sign "$=$," or an inequality like $<$ or $>$. Pronouns could be variables like x or y, $5x - 6$, x^2y, $8/x$. And they could be put together into a sentence like $5x + 14 = 22$. Both English and symbolic language are used in mathematics writing and mathematical lectures.

Summary

Recall that language is a type of abstraction (first order) used for communication. It is what enabled humans to pass their knowledge to the future generations and control the world. Although very useful, languages have their own limitations in that they can only furnish a finite number of names and words. Additionally, the ordinary languages are not designed for describing, expressing, or explaining the complex scientific ideas and concepts. The symbolic language of mathematics, on the other hand, is capable of expressing and communicating complex scientific ideas and relationships. Unlike short useable words, numbers have no limit. That is why we are better identified by a social security number than by a name.

References

Kenney, Joan M., Euthecia Hancewicz, Loretta Heuer, Diana Metsisto, and Cynthia L. Tuttle *Literacy Strategies for Improving Mathematics Instruction.* http://www.ascd.org/publications/books/105137/chapters/Mathematics-as-Language.aspx.

"The Language of Mathematics." https://www.mathsisfun.com/mathematics-language.html.

Wells, Charles. "The Symbolic Language of Math." http://abstractmath.org/MM/MMSymLang.htm.

Sports in United States and Europe

Sports are a popular pastime in most countries around the world. Other than certain globally popular sports, most regions of the world have their own preference in sports. Such sports are in a way reflection of the culture. In Europe, soccer is the most popular sport by a long shot. The second place is taken by ice hockey, because of their popularity in Scandinavian countries like Sweden and Norway. This is in sharp contrast to the USA, where football, basketball, and baseball are the big three sports. Some experts think that this is because Americans have shorter attention spans and prefer to play and watch higher-scoring games. The average attention span of an adult European ranges from fourteen to sixteen minutes; whereas for an American adult, it is eight to ten minutes. Because basketball and football lead to more scoring, they are more appealing to Americans. They also prefer to have side entertainment such as cheerleading and bands to increase excitement and to keep their attention.

In European football/soccer, the highest-scoring league is the German Bundesliga, according to British sports magazine *Mirror*. The Bundesliga matches average 2.93 goals per game, or about 1 goal per thirty minutes. Ice hockey's most prevalent European leagues are Sweden's SHL and Russia's KHL. In the first round of 2018 KHL playoffs, the number of goals per game was at 3.2, and in the SHL, the goals per game were 5.8. While this scoring average is higher than that of soccer, they are nothing in comparison to high-scoring sports such as basketball.

The lowest-scoring American sport is baseball. During the first four months of 2018, the MLB runs per game sat at 4.26, but the Yankees lead the league with 6.05. In the NFL 2017 season, points per game sat around 43, and in the NBA, the average points per game during the regular season was 213.2. To put that in perspective, 213. 2 points in sixty minutes is over 3.5 points per minute.

In the United States, football is known as the country's favorite sport. One hundred sixty million viewers worldwide watched the Super Bowl 48, with about 111.5 million of those views coming from the United States. That leaves 48.5 million viewers from other countries around the world (30.31 percent). While that sounds like a large number of viewers, it is nothing compared to what the FIFA World Cup tournament brings in. The 2014 FIFA World Cup Finals brought in a global audience of 1 billion who watched the game live and another 500 million watched the taped version.

Another difference of sports in the United States and Europe is the competitive balance. In the period of eleven years, starting 2006, there has been only three different teams that won the La Liga Division in Spain for football out of twenty. Realistically, if all twenty teams were evenly matched, there would be a 15 percent chance that one of those three teams would win each year. Over the course of eleven years, that would be an inconceivably small percent of what is actually happening. Similar things apply to the most popular league in the world—namely, English premiership. But since some teams such as FC Barcelona and Real Madrid are so dominant in this league, they take over the sport. In America's case, for the NFL, over the eleven-year period, nine different teams have won, and for the MLB, eight different teams have won. That would be a 6.25 percent chance of a team winning to championship titles during that stretch for the NFL and a 6.67 percent chance of an MLB team winning championships during that stretch. These numbers are much higher and more probable of happening than the football teams winning La Liga. In addition, the ironic part is that even though the odds of those three teams winning La Liga repeatedly are so small, it still happens each year. That is another huge difference between American and European sports.

References

Both, Andrew. "Super Bowl Has Ways to Go in Captivating Global Audience." *Reuters Sports News*, January 24, 2015. https://www.reuters.com/article/us-nfl-international/super-bowl-has-ways-to-go-in-captivating-global-audience-idUSKBN0KX0KK20150124.

Cox, Kevin. "Super Bowl 2014 Ratings Set New Record." CBS News, February 3, 2014. https://www.cbsnews.com/news/super-bowl-2014-ratings-set-new-record/.

Dubas-Fisher, David. "Is the Premier League REALLY the most entertaining league in the world?" *Mirror*, May 9, 2014. https://www.mirror.co.uk/sport/football/news/european-big-four-leagues-goals-3513388.

Elsworthy, Emma. *Average British Attention Span Is 14 Minutes, Research Finds.* December 28, 2017. Retrieved from independent.co.uk.

ESPN. *Scoring Average Per Professional Sports (2016–2017 seasons).* January 17, 2018. Retrieved from espn.com.

Kontinental Hockey League. https://en.khl.ru/.

"The Most Popular Sports in America." *Ranker.* https://www.ranker.com/crowdranked-list/most-popular-american-sports.

Official Site of Major League Baseball. https://www.mlb.com/.

Official Site of the National Football League. https://www.nfl.com/.

"One Billion Watched 2014 World Cup Final on TV." Reuters Sports News, December 16, 2015. https://uk.reuters.com/article/uk-soccer-world-television/one-billion-watched-2014-world-cup-final-on-tv-idUKKBN0TZ21Y20151216.

"Popular Sports in Europe." 7 Continents List, December 8, 2015. https://www.7continentslist.com/europe/popular-sports-in-europe.php.

Swedish Hockey League. https://www.shl.se/.

Sports and Math Education

This section presents a few math lessons using the links between mathematics, statistics, and sports. It provides a source of material that has an intrinsic interest level for most students. It shows just how mathematical and statistical analysis can be used to explain the structure of some sports-related events. Those who plan to become teachers, both at secondary and in higher education, will learn about a valuable source of material that will enrich their teaching of many mathematical topics. For the sporting enthusiast, this course demonstrates the effective mathematical and statistical model that can help to improve analysis of frequently used techniques. While modeling usually only confirms what intuition tells the expert sportsman, it is nevertheless gratifying to validate one's own interest.

Basketball Free Throws, a Teaching Tool

Basketball free throws can be used for teaching topics such as percentages, functions and their graphs, basic algebra (including inequalities), weighted averages, Simpson's paradox, elementary probability, basic geometry and trigonometry, geometric progression, and game theory. It also provides hints for how to use it for teaching more-advanced topics. In what follows, we demonstrate these together with their teaching objectives.

Percentages

Background information: How can we quantify the performances of basketball players attempting free throws so that we can compare them? Players are awarded different numbers of free throws based upon the plays in the game. The number of free throws the players make also differs.

Opening question

When in a basketball game a player gets fouled, more often than not, the announcers begin to talk about a player's free throw percentage, but what is percentage? How do we compute percentage?

One way to write a percentage is a decimal, and another is to look at it as a fraction. Percentage means out of one hundred, or out of the whole. For example, if during a game, a player makes five free throw shots out of ten attempted, we can say the following:

$$\text{Number of free-throws made} \Rightarrow \frac{5}{10} = \frac{x}{10}$$
$$\text{Number of free-throws attempted} \Rightarrow$$

Since percentages mean out of a hundred, we need to figure out what the x equals. If we multiply ten by ten, we get one hundred, so by multiplying the numerator by ten, we get fifty. So for that game, the player is a 50 percent free throw shooter. In other words, he makes approximately half of all free throws he attempts.

Activity

The table below shows the number of free throws a basketball player made during five different games.

	# Free Throw Attempts	# Free Throws Made
Game 1:	7	6
Game 2:	10	6
Game 3:	12	7
Game 4:	6	4
Game 5:	9	7

(a) From the table, figure out the player's percentage for each individual game (e.g., what was his percentage in game 1? game 2?) Then calculate his overall free throw percentage. Which games have a higher percentage than his overall free-throw percentage? Which ones have a lower one?

(b) Using one calculation from part (a), how did the player's overall free throw percentage change from game to game? Which game influenced his overall free throw percentage the most? The least?

Objective: Teaching of a function, graphing, and asymptotes as part of algebra related to computation of free throw percentages.

Opening question

A basketball player's free throw percentage tends to vary from game to game, since the free throws are dependent upon how many times a player is fouled. By using his previous free throw percentage, is there a way to calculate his current one? Can his future percentages be predicted? Can we then graph our prediction?

Activity

1. Suppose a basketball player is an 80 percent free throw shooter (he has made 80 percent of all total free throw attempts). During the last game, he made eight out of ten free throw shots. Is his percentage still 80 percent? Use algebra to show that this is the case. What happens if his free throw percentage was not 80 percent? Use a graph to present the result for all possible percentages.

Solution

To solve this problem, simple algebra can be used. Assume that before the last game, this player had made y out of x free throws attempted. The equation for his overall percentage prior to the last game is

$$\frac{y}{x} = 0.80 \quad or \quad y = 0.80x \tag{1}$$

Since tonight he has made eight out of ten, the equation for his free throw percentage P_N is now

$$P_N = \frac{y+8}{x+10} \tag{2}$$

Plugging $0.8x$ for y into equation (2), we obtain

$$P_N = \frac{y+8}{x+10} = \frac{0.80x+8}{x+10} = \frac{0.80(x+10)}{x+10} = 0.80$$

This shows that if a player makes the same percentage of free throws in just one game as his total percentage prior to that game, his percentage remains unchanged.

2. Suppose that in the last game played, a basketball player makes all ten out of his ten free throw shots. Prior to this game, he was an 80 percent free-throw shooter. Find an equation that would determine what his new free throw percentage is. Use y for the number of total free throws made and x for the total free throws taken.

Solution

$$\frac{y}{x} = 0.80 \quad or \quad y = 0.80x$$

$$P_N = \frac{y + 10}{x + 10} = \frac{0.80x + 10}{x + 10} = \frac{0.80x + 8 + 2}{x + 10} = \frac{0.80(x + 10) + 2}{x + 10}$$

$$= 0.80 + \frac{2}{x + 10}$$

Therefore, the increase will depend upon x.

From here, we can also predict what a player's new free throw percentage will be based upon the number of free throws he takes. The graph would look like something close to what is below.

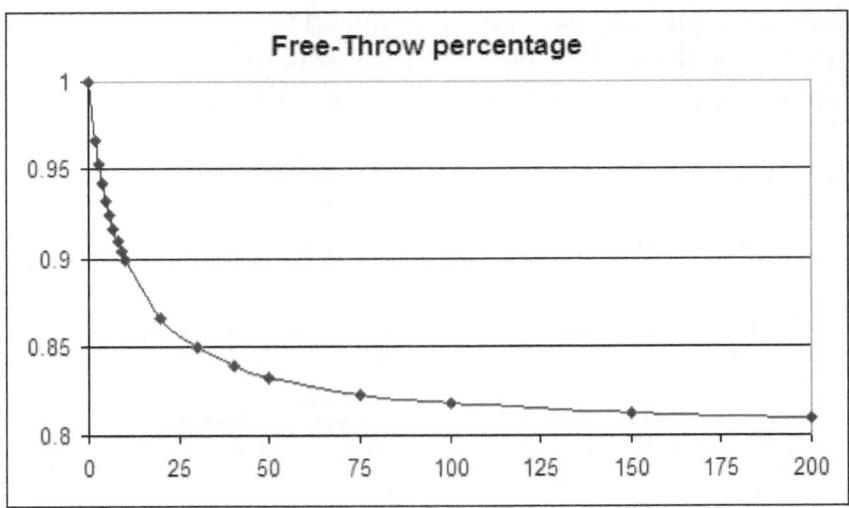

Objective: Teaching of limits as part of algebra related to computation of free throw percentages.

Opening question

How is the number of baskets attempted related to a basketball player's free throw percentage? Does the percentage fluctuate greatly, or does it approach some specific number?

Activity

Look at $P_N = .80 + \frac{2}{x+10}$ again. Try substituting a few different values for x, and make a table to show your results.

Examples: $x = 10 \quad \Rightarrow \quad P_N = .80 + \frac{2}{20} = .90$

$x = 90 \quad \Rightarrow \quad P_N = .80 + \frac{2}{100} = .82$

$x = 190 \quad \Rightarrow \quad P_N = .80 + \frac{2}{200} = .81$

What can we conclude from these examples? As x increases, the success of the last game becomes less significant. What number is the free throw percentage approaching?

1. Assume now that the player was a p percent free throw shooter for the first half of the season and a p' percent free throw shooter for the second half of the season. What is the player's percentage P_N for the nth season?

Solution

For the first half of the season, $\frac{y}{x} = p \quad or \quad y = px$

For the second half of the season, $\frac{y'}{x'} = p' \quad or \quad y' = p'x'$

$$P_N = \frac{y + y'}{x + x'} = \frac{px + p'x'}{x + x'} = \frac{px + px' + (p' - p)x'}{x + x} = \frac{p(x + x') + (p' - p)x'}{x + x'}$$

$$= p + (p' - p)\frac{x'}{x + x'} = \frac{x}{x + x'}p + \frac{x'}{x + x'}p' = p + (p' - p)a = (1 - a)p + ap'$$

where $\alpha = \frac{x'}{x + x'}$. Therefore, the new percentage is a weighted average of the old percentages.

Objective: Teaching ordinary and weighted averages as part of algebra related to computation of free throw percentages.

Opening question

How often does this player make his baskets? About how many does he make per game? Is this always a true number, or is it sometimes biased, either in support or opposition to the player?

Let's look at the probability of a player simply making one basket, and then making two consecutive baskets.

Examples:

$$x = x', \qquad \alpha = 1/2 \qquad \Rightarrow \qquad P_N = 1/2p + 1/2p'$$

$$x = 2x', \qquad \alpha = 1/3 \qquad \Rightarrow \qquad P_N = 2/3p + 1/3p'$$

$$x = 1/2x', \qquad \alpha = 2/3 \qquad \Rightarrow \qquad P_N = 1/3p + 2/3p'$$

.

.

.

$$x = kx', \qquad \alpha = 1/(1+k) \qquad \Rightarrow \qquad P_N = k/(1+k)p + 1/(1+k)p'$$

So, for example, if p = 0.50 and p' = 0.60, and k = 3, then

$$P_N = {}^3/_4\,(0.50) + {}^1/_4\,(0.60) = 0.525$$

Note that as $x \to \infty$, $P_N \to p$, and as $x' \to \infty$, $P_N \to p'$
If $p' > p$, then Pn > p
If $p' = p$, then Pn = p
If $p' < p$, then Pn < p

Now let's look at a player's average after two games. In the first game, he attempted ten free throws and made six of them. In the second game, he attempted fifteen and made twelve of them. What is his overall free throw percentage?

In game one, he made 6/10, which is 60 percent. In game two, he made 12/15, which is 80 percent. The most commonly made mistake is to add the two percentages and divide by two. To find the player's free throw percentage, one must take the total number of free throws made (i.e., 6+ 12 = 18) and divide that by the total number of baskets attempted (i.e., 10 + 15 = 25). So we get 18/25, or 72 percent, which is not equal to (60+80)/2=70 %. To see how we can find the overall average for the two games using the individual averages, we need to find the weighted average as follows:

$$\frac{10}{25}(0.60) + \frac{15}{25}(0.80) = \frac{6}{25} + \frac{12}{25} = \frac{18}{25} = 72\%$$

This is called weighted average. Here, the weights are (10/25) and (15/25). Note that the weights sum to one and are both nonnegative. In general, if we multiply each value with a nonnegative number and add, the resulting value is called a weighted average if the weights sum to one. In simple averages, the weights are all equal and are equal to $1/n$, where n is the total number of the measurements.

Activity

(a) Consider a player of your choice. Find his statistics (see for example nba.com) for two different periods of the regular season or consider regular season free throws versus the playoffs and carry on the necessary calculations.

(b) In eight games, a player makes fifty-nine free throws. The results of his games are given in the table below. Fill in all missing numbers of the table. If after game, his total number of baskets made increased to sixty-five, and if his overall free throw percentage is 81 and he attempted twelve free throws, how many baskets did he make in the sixth game?

Game	Numbers attempted	Numbers made	Free throw % (for specific game)
1	10		80
2	16	14	
3		17	85
4	12		75
5		12	80
6	12		

Bouncing Ball

The balls used in different sports have a different amount of bounce. Even the balls used in the same sport may bounce differently because of their age, coverings, or simply because they contain different amounts of air. For consistency, a standard for bounciness must be established for the ball in each sport.

On way to measure the bounciness of a ball is through a quantity known as the coefficient of restitution (COR). The COR is defined as square root of the ratio of the rebound height to the initial height from which the ball was dropped:

$$COR = \left(H_{Rebound}/H_{Initial}\right)^{1/2}$$

Calculation

Provide answers to the following questions:

1. A tennis ball has a COR of 0.53. If this ball is dropped from a balcony at a height of eight feet, how high will the ball bounce?
2. From what height does a tennis ball with a COR of 0.54 should be dropped to bounce twelve feet?

3. Suppose that a tennis ball with a COR of 0.55 is thrown at a wall with a speed of 65 mph. With what speed will it rebound?

Critical thinking

Make up a question pertaining to this lesson that you, the student, would ask if you were a teacher. For example:

- You may want to know if bounciness can be quantified differently, and if so, what would a consequence of that be?
- Is COR independent of speed? That is, if you throw two identical balls at a wall, one with the speed twice the other, would speeds of the rebounds be two to one too?

Estimation and modeling

Select a tennis ball. Suppose that the ball is dropped from a point h feet high and has a COR equal to c.

(a) Develop a model to calculate the heights of the first, second, third, and other bounces.

(b) Think about the total distance traveled by the ball after one, two, etc., bounces. See if you could develop a model for this. Use the models you developed for a long-run prediction. For example, what is the estimate for total distance traveled after n bounces?

(c) Suppose that c is unknown. How do you proceed to estimate that? For example, you may estimate COR by measuring the bounce n times and by averaging the results. Suggest a value for n.

(d) Estimate the second bounce by first directly measuring it n times and averaging as in part (c). Then apply the mathematical model in part (a) and predict the average of the second bounce. Which estimate do you prefer?

(e) Repeat part (*b*) for the third, fourth, and other bounces. Do you see any pattern?

Trajectories, Some Classical Functions, and Modeling Building

Suppose now that a ball will be hit by a racket at a certain angle (similar to serves in tennis). After hitting the ground, it will bounce and follow a path like parabola (similar to the trajectory of, for example, a lob shot in tennis). Model this scenario first for a fixed angle and fixed initial speed. Repeat the process by keeping one variable fixed and letting the other vary. Finally, let both variables vary.

Pythagorean Theorem and Baseball

The original Pythagorean theorem of baseball was devised by Bill James in the 1980s. The "theorem" predicts the winning percentage of a baseball team based on how many runs the team scores as well as how many runs it allows. When a team scores fewer runs than it allows, the model suggests that the team should have a losing record, and when a team scores more runs than allowed, they should have a winning record. In the 2001 season, the New York Mets allowed more runs than they scored but still had a winning record, so they were considered an overachieving team. The Colorado Rockies, on the other hand, scored more runs that they were allowed but still had a losing record, so they were considered an underachieving team.

Now, about twenty years later, Michael Jones and Linda Tappin of Montclair State University in New Jersey have devised mathematically simpler alternatives to the Pythagorean theorem of baseball.

(a) To predict the winning percentage of a team, one new

model simply uses addition, subtraction, and multiplication. It starts with the total runs scored by the team in all of its games (R_S) and subtracts the runs that it allows (R_A) and then multiplies it by a number Beta (β), which is chosen to produce the best results. For the 1969–2003 seasons, the optimal values of B range from 0.00053 to 0.00078 with an average of 0.00065.

Adding 0.5 to the result gives the predicted winning percentage of the team. The resulting linear formula looks like this:

$$P = 0.5 + \beta(R_S - R_A)$$

This equation can be used to teach simple linear regression. If an equation takes the form $y = b + mx$, then here we are using x to predict y, and the equation is said to be linear. In the above equation using the baseball variables, the R_S and R_A are the values that are used in trying to predict what P, the winning percentage, will be. One advantage of using this formula is we can find a range for P.

(b) In contrast, the original Pythagorean Theorem of baseball is a little more complex. It uses the square of the previously used variables. The resulting formula is

$$P = R_S^2 / (R_A^2 + R_S^2).$$

This equation gets its name because of its similarity to the Pythagorean theorem in geometry, which relates the lengths of the sides of a right triangle as $a^2 + b^2 = c^2$, where a and b are the shorter sides of the triangle and c is the hypotenuse.

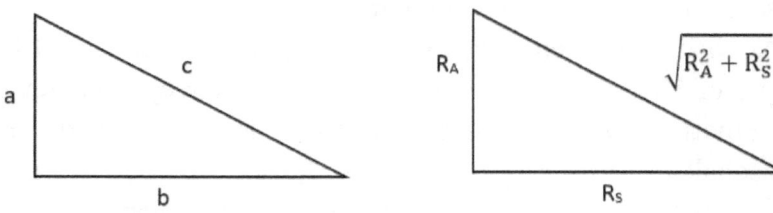

Examples using both methods (a) and (b):

1. In the 2003 season, the Philadelphia Phillies scored 791 runs and allowed 697 runs to be scored against them. According to the 2003 MLB relative power index, the winning percentage (PCT) for the Phillies was 0.531. Test to see how close these methods predict the actual PCT.

   (a) The range of values for the winning percentage can be predicted for the different values for Beta (β).

For $\beta = 0.00053$, $P = 0.5 + (0.00053)(791 - 697) = 0.54982$
For $\beta = 0.00078$, $P = 0.5 + (0.00078)(791 - 697) = 0.57332$

Therefore, using the linear method, it could be predicted that the Phillies had a winning percentage between 0.54982 and 0.57332.

   (b) Using the Pythagorean theorem,

$$P = (791)^2 / [(697)^2 + (791)^2] = 0.56292$$

In this case, both strategies predict a very similar result.

2. The Baltimore Orioles scored 743 runs in the 2003 season, while allowing 820 to be scored upon them. According to the 2003 MLB relative power index, the winning percentage (PCT) for the Orioles was 0.490. Test to see how close these methods predict the actual PCT.

   (a) The range of values for the winning percentage can be predicted for the different values for Beta (β).

For $\beta = 0.00053$, $P = 0.5 + (0.00053)(743 - 820) = 0.45919$
For $\beta = 0.00078$, $P = 0.5 + (0.00078)(743 - 820) = 0.43994$

Therefore, using the linear method, it could be predicted that the Orioles had a winning percentage between 0.43994 and 0.45919.

(b) Using the Pythagorean Theorem,

$$P = (743)^2 / [(820)^2 + (743)^2] = 0.45085$$

In this case, both strategies again predict a very similar result.

The question is whether the Pythagorean Theorem was needlessly more complicated. It was found that for the baseball seasons from 1969 to 2003, the new linear formula predicts the percentages almost as well as the old theorem. The one real exception was the 1981 season in which there was a baseball strike. Therefore, the new linear method may be a better and simpler solution to predicting the winning percentages of baseball teams.

Penalty Kicks

A penalty kick (penalty) is a free kick taken from the penalty mark thirty-six feet (twelve yards or approximately eleven meters) from the center of the goal and with no player other than the goalkeeper of the defending team between the penalty taker and the goal.

Penalty kicks occur during a normal play. They also occur in some tournaments to determine who progresses after a tie game; though similar in procedure, these are not penalty kicks and are governed by slightly different rules.

In practice, penalties are converted to goals more often than not, even against world-class goalkeepers.

A penalty kick may be awarded when a defending player commits a foul punishable by a direct free kick against an opponent or a handball, within the penalty area (commonly known as "the box" or "18-yard box"). It is the location of the offense and not the position of the ball that defines whether a foul is punishable by a penalty kick or direct free kick, provided the ball is in play. The penalty taker (who does not have to be the player who was fouled) must be clearly identified to the referee.

The penalty kick is a form of direct free kick–meaning, a goal may be scored directly from it. If a goal is not scored, play continues as usual. As with all free kicks, the kicker may not play the ball a second time until it has been touched by another player even if the ball rebounds from the posts. However, a penalty kick is unusual in that, unlike general play, external interference directly after the kick has been taken may result in the kick being retaken rather than the usual dropped ball.

Basic Geometry and Trigonometry

Background information: Penalty kicks are normally taken from the penalty mark, which is a midline spot thirty-six feet from center of the goal. The penalty mark has the same distance from both goalposts.

Opening question: Suppose penalty kick can be taken from any point in the field as long as its distance from the center of goal is thirty-six feet. Suppose also that the goalkeeper stands in the center of the goal and can protect the area inside a circle of radius eight feet. From which point on the penalty taker has the greatest chance of scoring?

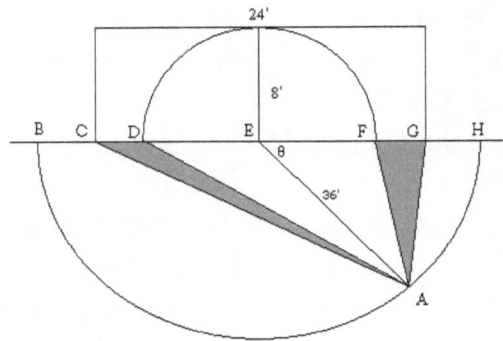

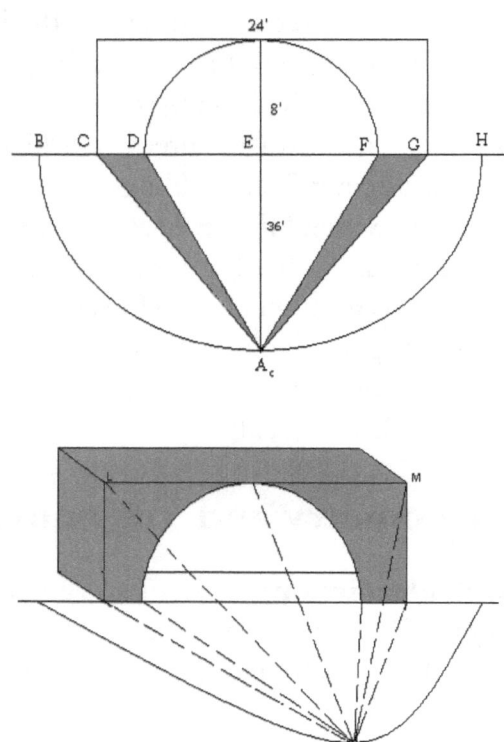

Solution

Consider a kick taken from an arbitrary point A shown in figure 1. The goal width (CG) is 24 feet or 7.32 meters. Since we assume that the goalkeeper stands at the center of the goal and can protect 8 feet or 2.44 meters (r) to his either side (segment DF), a goal can be scored only if the penalty taker kicks the ball inside the triangles ACD or AFG.

Let θ denote the angle $\angle AEF$ and assume, for now, that the penalty taker can only kick the ball on the ground. Then it is easy to show that the areas of triangles AEF and AEG are respectively 36 x 8 x Sin(θ)/2 ft (11 x 2.44 Sin(θ))/2 m) and 36 x 12 Sin(θ)/2 (11 x 3.66 Sin(θ))/2 m). Note the same values present the areas of the triangles ADE and ACE, respectively. Using these, the total area inside the triangles ACD and AFG is 36 x 2 Sin(θ) (11 x 0.61

Sin(θ)m). To determine the best point, we note that since sin(θ) is an increasing function for $0 \leq \theta \leq \pi/2$, the total area increase penalty taker moves the ball toward penalty mark–that is, the center of the half circle. In fact, the total area is maximum when the ball is moved to the penalty mark where $\theta = \varpi/2$.

Now, for the general case, the penalty taker can kick the ball inside the volume shown in figure 1. Since for any point selected on the half circle the base of this volume is the same, the choice with the largest possible height produces the maximum volume. This, referring to the case discussed above, shows that, as before, the penalty mark is the best choice.

Activity

- Determine what happens to the effective scoring area if the penalty can be taken from a point with distance less than thirty-six feet from the center of the goal; the goalkeeper's side-to-side range changes; or the goalkeeper moves forward out of the goal.
- For penalty kick taken from the penalty mark, find the effective scoring angle (not area).
- Suppose that the penalty taker can kick the ball with the speed of v. Could you think of a way of incorporating this dimension of the penalty kick into the problem? Would this change your answer to the problem of choosing the point with the highest chance of scoring?

Game Theory

Background information: Consider the game played between the opposing goalie and a penalty taker. The penalty taker can elect to kick toward one of the two goalposts, or else aim for the center of the goal. The goalie can decide to commit in advance (before the kick) to either one of the sides, or else remain in the center until he sees the direction of the kick.

Opening question: Can you think of optimal strategies for the penalty taker and goalie?

Solution

This two-person zero-sum game can be represented as follows, where the payoffs are the probability of scoring a goal.

		Goalie		
		Breaks Left	Remains Center	Breaks Right
	Kicks Left	0.5	0.9	0.9
Kicker	Kicks Center	0.1	0	0.1
	Kicks Right	0.9	0.9	0.5

If we assume that decisions between the left or right side are made symmetrically (with equal probabilities), then this game can be represented by a 2×2 matrix, where $0.7 = (1/2)(0.5) + (1/2)(0.9)$:

		Goalie	
		Remains Center	Breaks side
Kicker	Kicks center	0	1
	Kicks side	0.9	0.7

The optimal strategies of the kicker and goalie are to

    a. more often kick center and remain center;
    b. more often kick side and break side; and
    c. choose center and side equally often.

Activity

(a) Suppose that a penalty taker kicks rights more often than left. Discuss how that may change the optimal strategies discussed above.

(b) Make any changes you consider appropriate or interesting and discuss their consequences.

Closing remarks

How do you summarize the penalty statistics for players, teams, and tournaments? This can be used to teach descriptive statistics.

How do you compare players and teams? This can be used to teach performance measures, measures of relative standing, z-score, etc.

A claim is made about a player. Using his/her statistics, how do you validate the claim? This can be used to teach hypothesis testing.

Using the past statistics of a player, how do you predict his/her future performance? This can be used to teach estimation (prediction), confidence intervals, regression, time series, and forecasting.

Suppose that the outcomes of successive penalty kicks have dependency. How do you model and analyze such data? This can be used to teach Markov chain.

Geometric Progression

Two friends (A and B) are about to go out for lunch. To add extra excitement to this, A suggests playing the following game to see who should pay for lunch. "Let's try free throws in turn. Whoever makes the first free throw wins. The loser will have to pay for lunch." Also, to encourage B to accept, she lets B go first.

(a) Assume that both A and B are 50 percent free throw shooters.
1. Find the probability that A has to pay for lunch.
2. Find the probability that B has to pay for lunch.

(b) Answer the questions in part (a) assuming that both players are
1. p percent shooters; and
2. A is a p_A free-throw shooter and B is a p_B percent free-throw shooter.

(c) Add any further analysis you could think of (one or two).
(d) Help a math teacher to use this for a lesson plan of a topic of your choice.

Solution: Assume that the results of successive free throws are independent.

Start

(a) $P(A \text{ pays}) = \dfrac{1}{2} + \left(\dfrac{1}{2}\right)^3 + \left(\dfrac{1}{2}\right)^5 + \cdots = \dfrac{1/2}{1-(1/2)^2} = \dfrac{2}{3}$

$P(B \text{ pays}) = 1 - \dfrac{2}{3} = \dfrac{1}{3} = \left(\dfrac{1}{2}\right)^2 + \left(\dfrac{1}{2}\right)^4 + \left(\dfrac{1}{2}\right)^6 + \cdots = \dfrac{(1/2)^2}{1-(1/2)^2} = \dfrac{1}{3}$

(b)

1. $P(A \text{ pays}) = p + (1\text{-}p)(1\text{-}p)p + (1\text{-}p)(1\text{-}p)(1\text{-}p)(1\text{-}p)p + \ldots = p + (1\text{-}p)^2 p +$

$(1\text{-}p)^4 p + \ldots = \dfrac{p}{1-(1-p)^2} = q_A$

$P(B \text{ pays}) = 1 - \dfrac{p}{1-(1-p)^2} = \dfrac{1-p-(1-p)^2}{1-(1-p)^2}$

2. $P(A \text{ pays}) = p_B + (1\text{-}p_B)(1\text{-}p_A)p_B + (1\text{-}p_B)(1\text{-}p_A)(1\text{-}p_B)(1\text{-}p_A)p_B + \ldots$

$\qquad\qquad\quad = p_B + (1\text{-}p_B)(1\text{-}p_A)p_B + (1\text{-}p_B)^2 (1\text{-}p_A)^2 p_B + \ldots$

$\qquad\qquad\quad = \dfrac{p_B}{1-(1-p_A)(1-p_B)}$

$P(B \text{ pays}) = 1 - \dfrac{p_B}{1-(1-p_A)(1-p_B)} = \dfrac{1-p_B-(1-p_A)(1-p_B)}{1-(1-p_A)(1-p_B)}$

(c)

- Let us assume that $P(A \text{ pays}) = q_A$. Then we can use smarter mathematics as follows:

$$q_A = p_B + (1- p_B)(1- p_A)q_A \Rightarrow q_A = \frac{p_B}{1-(1- p_B)(1- p_A)}$$

This is because either B makes it in the first attempt or she misses and A misses too, in which case the game starts all over again.

- When $p_A = p_B = p$, show that $P(A \text{ pays}) > P(B \text{ pays})$–that is, the game is not a fair game.
- When $p_A \neq p_B$, the game is fair when $q_A = q_B$. That is,

$$\frac{2p_B}{1-(1- p_A)(1- p_B)} = 1 \Rightarrow p_A - p_B - p_A p_B = 0 \Rightarrow p_A = \frac{p_B}{1- p_B}, \left(p_B < \frac{1}{2}\right)$$

For example, $p_B = \dfrac{1}{3} \Rightarrow p_A = \dfrac{1}{2}$

- Add a third person.
- Try k free throws. First person to make h wins.

- Each person tries free throws until she makes r. Anyone who does in a smaller number of attempts wins.

(d) A teacher can use this for lesson plans on
   (1) geometric progression: part: (a) and (b)
   (2) univariate and bivariate function: part (b)
   (3) graphing a function: part (b)
   (4) teaching inequalities: part (c)
   (5) fair games: part (c)

Simpson's Paradox

Consider two players A and B. Suppose that in the first half of the season, player A made only 20 of his 100 free throw attempts. In the same period, the player B made 50 of his 210 free throws. So in the first half of the season, B performed better (24 percent versus 20 percent). Suppose that in the second half of the season, A made 40 of his 60 and B made 15 of his 20 attempts, respectively. Thus, again, B performed better (75 percent versus 67 percent). Now, if we combined these statistics for the season, we have $60/160 = .38$ for A and $65/230 = .28$ for B. That is, A did better for the season. This means that although B performed better in both the first and second half of the season, his performance was worse than A in the whole season. This is, of course, a paradox known as Simpson's paradox. The table below summarizes these statistics.

	First Half	Second Half	Season
Player A	$20/100 = 0.20$	$40/60 = 0.67$	$60/160 = 0.38$
Player B	$50/210 = 0.24$	$15/20 = 0.75$	$65/230 = 0.28$

Now, it is clear from this summary that different conclusions may be drawn from looking at whole or parts of players' statistics. The lesson here is that we must be extremely careful in reporting averages (percentages) unless the groups under consideration are fairly homogeneous.

A similar example: During the last basketball season, Jim attempted one hundred free throws in the first half of the season and made thirty. So his free throw percentage was $30/100 = 0.300$. He also attempted twenty in the second half of the season and made eight ($8/20 = 0.400$). His stat was better than Curt's stat, $5/20$ (0.250) for the first half of the season and $35/100$ (0.350) for the second half of the season. For the season, however, Curt's free throw percentage, $(5+35)/(20+100) = 40/120 = 0.333$, was higher than Jim's $(30+8)/(100+2) = 38/120 = 0.317$. So who do you think was a better free-throw shooter in that season? Here, mathematics helps you to decide. How? The conclusion based on detail should always be preferred. Two dimensions versus one, so Jim.

Baseball Rookie: Old-Timer Paradox

I thought rookie was better for sure, but little did I know.

Even a curiosity glance at newspapers shows the extent to which use of classical statistical methods has become a part of everyday life. This is particularly the case for popular sports such as baseball. Both experts and fans analyze the data to draw conclusions about players and teams. But do application of popular classical statistical methods always lead to an appropriate conclusion? Here, I want to present an instance in which different analyses of same data lead to apparently paradoxical results. Hopefully, it would not come up in the new season.

We describe an example of the so-called Simpson's paradox, named after a statistician who gave a careful discussion of it. Consider a scenario involving a rookie, R, who is trying to break into a baseball lineup by replacing an old-timer, O. He is told by the manager that the decision will be made on the basis of hitting ability, since he and O are equally proficient as fielders. The rookie feels assured that he will start because he is batting 0.224 and O is batting only 0.186 (both are good field, poor hit, players.) However, the rookie is dismayed to learn that O is designated to start the first game. He asks the manager why he is not starting, since his average is better than that of O. "Well," explains the manager, "the opponents are using a right-handed pitcher today, and O has a better average than you against right-handed pitches. His average is 0.179 and yours is 0.162."

Satisfied but chagrined by the explanation, R waits until the opponent's pitcher is left-handed, for then, surely, he will get to start. But that day comes, and O is once again the starting player. The manager points out that "O has a better average than you against left-handed pitchers, too, batting 0.560 compared to your 0.332."

The rookie is no longer satisfied with the explanation, and he asks, "How can O be better than me against both left- and right-handed pitchers but poorer overall?"

The data shown in table below are persuasive. The arithmetic is correct, and careful analysis of the table may reveal the source of the paradox. The total batting percentage is determined by the percentages against the two types of pitchers but is determined differently for each of the two players. The overall percentage is a *weighted average* of the individual percentages, and the weights are determined by the number of times at bat against each of the types of pitchers. Although we expect the weighted average to preserve the order of the individual percentages, it does not.

	Times at Bat		Hit	Average
Totals	O	4766	888	0.186
	R	1276	186	0.224
Against right-	O	4675	837	0.179
handed pitchers	R	809	131	0.162
Against left-	O	91	51	0.560
handed pitchers	R	467	155	0.332

Although the story of the rookie is fictitious, the data is not. The lesson here is that we must be careful in reporting averages unless the groups under consideration are fairly homogeneous. In other words, one cannot always trust conclusions drawn from marginal tables, and one way to avoid that is to consider all the dimensions of a table.

A different version of this counterintuitive puzzler was adapted from a *Car Talk* radio show by David Lincoln, the Catchup Math account manager:

Two baseball rookies had a bet about their first season: the one with the highest batting average would win. They both had the same number of at bats. Player A bested player B in the first half of the season (0.300 versus 0.250) and also in the second half (0.400 versus 0.350). And yet player B won the bet! How could this be possible!

One answer: Player A was 30/100 (0.300) in the first half and 8/20 (0.400) in the second half. Player B was 5/20 (0.250) in the first half and 35/100 (0.350) in the second half. Thus, player B's average of 40/120 (0.333) was higher than player A's average of 38/120 (0.317).

As mentioned earlier, these are examples of what is known as Simpsons' paradox. The paradox occurs because collapsing the data can lead to an inappropriate weighting of the different populations. The lesson here is that one must be extremely careful in reporting averages unless the groups under consideration are fairly homogeneous. In general, the analysis based on details (higher dimensions) are preferred.

References

Smith, Arthur. "At the Plate, a Statistical Puzzler: Understanding Simpson's Paradox." *The State of the USA | At the Plate, a Statistical Puzzler: Understanding Simpson's Paradox*, August 20, 2010, www.stateoftheusa.org/content/at-the-plate-a-statistical-puz.php.

Hot Hand in Sport

"Hot hand" in basketball is a phenomenon that most basketball fans accept and consider well understood. It is a topic that has generated a great deal of interest among researchers in areas such as psychology, statistics, and physical education. For a player who is hot, basket appears as if it is so wide that he/she will make shot after shot with ease. The opposite situation may be called "cold hand."

As mentioned, several publications have demonstrated that statistically hot hand phenomenon may not even exist. Amos Tversky, a psychologist who studied every basket made by the Philadelphia 76ers for more than a season, has concluded the following:

1. The probability of making a second basket (hit) did not rise following a successful shot (success did not breed success).
2. The number of runs or baskets in succession was no greater than what a coin-tossing model would predict.

To clarify, consider a good player who makes half of his shots. For such a player, four hits in a row (HHHH) is expected to occur once in sixteen sequences of four shots, as this is one of the sixteen equally likely outcomes. The chance that this player hits at least three shots in a row (HHHH, HHHM, MHHH) is 3/16, assuming that trials are independent. Does this mean that the player has a hot hand? To add to this, suppose that I flip a coin four times and get four heads in a row. Could I claim that I have a hot hand, say, for heads?

Clearly, a great player will have more sequences of five hits than an average player but not because he has greater will or gets in that magic rhythm more often. He/she has longer run because his average success rate is so much higher and has a much better chance of having more frequent and longer sequences. Suppose that we simulate a great player's game whose chance of making a field goal is 0.6. Such a player will get five in a row about once in every thirteen sequences since the chance of this event is 1/13 or $1/(0.6)^5$. If another player's chance of making a field goals is 0.3, he will get his five hits in a row only about once in 412 times. In other words, we need no special explanation for the apparent pattern of long runs, except perhaps when the performance is far beyond what is expected from a player based on his past statistics.

Similar studies in baseball indicate that nothing ever happened in baseball above and beyond the frequency predicted by fair or loaded coin-tossing models. The longest run of wins or losses are as long as they should be and occur about as often as they ought to. Again, this may also be demonstrated using computer simulation. However, this rule has one exception that should never have occurred: DiMaggio's fifty-six-game hitting streak in 1941. Purcell calculated that to make it likely (probability greater than 50 percent) that a run of even fifty games will occur once in the history of baseball up to now (and fifty-six is a lot more than fifty in this kind of league), baseball's rosters would have to include either four lifetime 0.400 batters or fifty-two lifetime 0.350 batters over careers of one thousand games. In actuality, only three men have lifetime batting averages in excess of 0.350, and no one is anywhere near 0.400. He then concluded that DiMaggio's streak is the most extraordinary thing that ever happened in American sports.

Having stated all of the above, the questions still remain: how do people define hot hand, and are their definitions based on patterns, reality, or perceptions? A study by Larkey et al. (1989) reveals that different measures for hot hand leads to consider different players hot. It is interesting to note that, although the precise meaning of a term like *hot hand* is unclear, its common use implies a shooting record that departs from coin tossing with the probability of success greater than that expected. An equally interesting point relates to the fact that fans who talk about hot hand usually refer to patterns of streak shooting, or something that is noticeable, memorable, or unlikely. Examples are observations such as HHHHHM or MHHHHH. But then, why is an outcome like HMHMHM, which presents a sequence of hits followed by misses and vice versa, not considered a notable or extraordinary pattern, and why is it nameless? As is pointed out in *The Skeptic's Dictionary,* the clustering illusion is the intuition that random events that occur in clusters are not really random events. The illusion is due to selective thinking based on a false assumption. A good example occurs in the lottery. People think that a number like 2145362 is more random than 2222222; whereas when choosing a seven-digit number randomly, they have the same chance of being selected.

Here is another example supporting the fact people only recognize certain patterns and ignore many others. Let us replace hit by 1 and miss by 0. Then we have a sequence of ones and zeros (a number in binary system). Consider, for example:

$$0\ 0\ 0\ 0\ 0\ 0\ 1\ 1\ 1\ 1\ 1 \text{ and } 1\ 1\ 1\ 1\ 0\ 0\ 1\ 1\ 0\ 1\ 0.$$

The first sequence may be recognized as having a pattern. But what about the second sequence? In fact, this is a famous pattern. It is the binary presentation of the number 1946, my birth year. To me, this is a recognizable pattern, but to the others it is not. A similar example happens in poker. Getting a royal flush is surprising even though the chance of any hand in poker, with no particular name, is extremely low. It is also possible to argue in favor of hot hand. In an interesting paper, Wardrop (1995) presents many discussions concerning an inherent weakness in the methods used by Gilovich et al. (1985) and Tversky and Gilovich (1989). Hooke (1989) discusses the inherent difficulty of using statistical methods to study complete phenomenon such as a game of basketball. Hale (1999) has discussed this issue and has raised several questions. He has argued that hot hand is an internal phenomenon and that the sense of being "hot" does not predict hits or misses. When a player realizes he is hot, he tends to push the envelope and attempt more difficult shots, which then leads to predictably a failure. He might also raise the question: precisely how unlikely does a streak of success need to be before we are prepared to count it as a legitimate instance of hot hand?

According to Hale, there are three prominent arguments that conclude there are no hot hands in sports. The first argument of the hot hands critics creates a tradition in the very act of destroying it. By making "success breeds success" a necessary condition of having hot hands, the critics have established a previously undefended and barely articulated account of hot hands, only to demolish it. Instead, he has argued that there are good reasons to reject "success breeds success" as a requirement for having hot hands. While it is true that many players believe that future success is more likely when they are already hot, either this is only a belief that their current state has causal efficacy into the future, or is its inductive reasoning that their current high rate of success is evidence of future success? Yet neither disjunction makes "success breeds success" part of the concept of having hot hands.

The next two arguments offered by the critics of the hot hand theory are of a well-known skeptical pattern: set the standards for knowledge of something extremely high, and then show that no one meets those standards. The canonical reply to this strategy, of which he availed himself, is to reject those standards in favor of more modest ones that charitably preserve our claims of knowledge. The skeptical insistence upon exceedingly rare streaks or statistically remote numbers of streaks as being the only legitimate instances of hot hands is arbitrary and severe. He has then argued that "being hot" denotes a continuum, one that is nothing other than deviation from the mean itself. This obviously comes in degrees.

To summarize the arguments against the hot hand, Tversky and Gilovich, using several data sets, concluded that existing data does not support "hot hand." They have devised a clever experiment to obtain convincing evidence that knowledgeable basketball fans are much too ready to detect occurrences of streak shooting and the "hot hand" in sequences that are, in fact, the outcome of Bernoulli trials. To clarify this argument further, they have considered the data concerning the free throws for nine regular players of Boston Celtics from 1980 to 1982. Then they asked the following question: when shooting free throws, does a player have a better chance of making the second shot after making the first shot than after missing the first shot? To answer, they have chosen a sample of one hundred Cornell and Stanford students randomly. The responses were 68 percent yes implying hot hand and 32 percent no implying independence or negative association.

After analyzing the data, Tversky and Gilovich concluded that the data provided no evidence that the outcome of the second shot depends on the outcome of the first one. Adams (1992), using data on eighty-three players, showed that the mean interval from making a field goal (n = 372) to making a field goal in nineteen NBA games did not differ from the mean interval from making to missing (n = 394), which further challenges assumptions regarding hot hand.

In a more recent study, Koehler and Conley, in "The Hot Hand Myth in Professional Basketball," which appeared in the *Journal of Sport and Exercise Psychology,* have offered further evidence against hot hand in a unique setting, the NBA Long Distance Shootout contest. They have concluded that declarations of hotness in basketball are best viewed as historical commentary rather than as a prophecy about future performance.

Having discussed opposing views, the final and perhaps more important question is why arguments for and against hot hand seem convincing. In an attempt to answer this question, Wardrop (1995) performs a very interesting analysis of data for Boston Celtics players. His analysis is based on, as he stated, the fact that the data available to laypersons may be very different from the data available to professional researchers. In addition, laypersons unfamiliar with a counterintuitive result, such as Simpson's Paradox, may give the wrong interpretation to the pattern in their data and their analysis. There are many problems of this type in probability theory in which the right answer is counterintuitive. Here, we will borrow the data and its analysis presented in Wardrop for demonstration. To understand Wardrop's analysis, consider the following data presented in Wardrop (1995). Observed frequencies for pairs of free throws by Larry Bird and Rick Robey and the collapsed table (Wardrop 1995, table 1).

Larry Bird			
	Second		
First	Hit	Miss	Total
Hit	251	34	285
Miss	48	5	53
Total	299	39	338
Rick Robey			
	Second		
First	Hit	Miss	Total
Hit	54	37	91
Miss	49	31	80
Total	103	68	171
Collapsed Table			
	Second		
First	Hit	Miss	Total
Hit	305	71	376
Miss	97	36	133
Total	402	107	509

Let Ph = the proportion of first shot hits that are followed by a hit and Pm, the proportion of first shot misses that are followed by a hit.

Ph = 251/285 = 0.881 and pm = 48/53 = 0.996 for Larry Bird,
Ph = 54/91 = 0.593 and pm = 49/80 = 0.612 for Rick Robey.
But for collapsed data, we have

Ph $= 305/376 = 0.811$ ($= (285/376)(251/285) + (91/376)$ $(54/91)$), and

Pm $= 97/133 = 0.729$.

Note that, contrary to the hot hand theory in the sense of success breeds success, each player's shot is slightly better after a miss than after a hit. Although, as is shown by Wardrop, the differences are not statistically significant.

It is possible, of course, to ignore the identity of the player attempting the shots and examine the collapsed data. For example, on 509 occasions, either Bird or Robey attempted two free throws; on 305 of those occasions, both shots were hit; and so on. For the collapsed data, ph $= 0.811$ and pm $= 0.729$. These values support the hot hand theory—that is, a hit was much more likely than a miss to be followed by a hit.

The data from Bird and Robey illustrate Simpson's paradox—namely, ph $<$ pm in each component table, but ph $>$ pm in the collapsed table. It is easy to verify algebraically that the proportion of successes for a collapsed table proportions equals the weighted average of individual player's proportions, with weights equal to the proportion of data in the collapsed table that comes from the player. For the after-a-hit condition, for example, the weight for Bird is $285/376 = 0.758$, the weight for Robey is $91/376 = 0.242$, and the proportion of successes for the collapsed table, $305/376$ $= 0.811$, is $(285/376)(251/285)=(91/376)(54/91)$. As a result, even though both Bird and Robey shot better after a miss than after a hit, the collapsed values show the reverse pattern because of the huge variation in weights associated with each player. In short, Wardrop has concluded that Simpson's paradox has occurred because the after-a-miss condition, when compared to the after-a-hit condition, has a disproportionately large share of its data originating from the far inferior shooter Robey.

Patterns That Do Not Exist

Athletes, coaches, and fans of sports such as basketball and baseball overwhelmingly believe in the "hot hand," the idea that a player whose shooting percentage is higher than normal is likely to keep shooting better than normal—at least for a while. Indeed, coaches often rotate players based on a sense of who is hot or cold, even though academic scholars thought they had debunked this idea years ago. Starting with a famous 1985 study of basketball shooting, experts have argued in dozens of papers that hot hand is nothing more than a random occurrence of statistical noise. In "Investment Bankers Can Learn a Thing or Two from Athletes on a 'Hot Streak,'" Edmund Andrews argues that even though athletes seem to go on hot streaks, these are just random fluctuations without predictive value. People who believe in hot hands, the skeptics argue, are seeing patterns that do not exist.

In a major new study of baseball data, Jeffrey Zwiebel at Stanford and Brett Green at the University of California, Berkeley, concluded that hot hands are real and have predictive power. We may wonder why these business professors would jump into a sports debate, but the "hot hand fallacy" is used by supporters of behavioral economics to argue that people can be irrational. For example, behaviorists have argued that investors often get lured into bad decisions by seeing patterns that are not real. According to behaviorists, investors make a wide range of cognitive mistakes that range from overconfidence in their own abilities to a tendency to overreact to the news.

According to Andrews, that critique gained a lot of traction in the wake of the mortgage bust and the great financial crisis. A 2009 publication lays out the theory and potential financial applications of the hot hand fallacy. A 2012 paper argues that the hot hand fallacy explains why people pay for useless investment advice. In a new publication, German scholars even cite evidence that people who believe in the hot hand fallacy are more at risk of long-term unemployment.

Andrews says that the hot hand fallacy has its roots at Stanford. Thomas Gilovich, a graduate student in the early 1980s, began comparing the widespread perception of hot streaks in basketball to the hard data. Gilovich led a study showing that the hot hand did not really exist; the shooting records of the Philadelphia 76ers provided no predictive value of subsequent shots. A player might be hot one minute but not the next. Fans and even sports professionals, they concluded, were making decisions based on myopic impressions.

Zwiebel and Green argue that the original finding failed to account for a key issue. In basketball, the opposing team quickly adjusts to a hot player, devoting extra coverage and forcing that player to attempt more difficult shots. As a result, it's inevitable that a hot player's shooting percentage will decline—not because the hot streak was an illusion, but rather that the hot player attracted more opposition.

They also point out that baseball is different because pitchers and coaches have limited ability to redeploy resources against a hitter on a hot streak. Pitchers and coaches play to a hitter's particular weaknesses, but they cannot put more people in the hitter's way.

To test their ideas, Zwiebel and Green amassed data on two million Major League Baseball at bats over twelve years. They looked at ten categories of performance, from batting averages and home-run percentages to strikeout rates. For pitchers, they looked at data such as the average number of hits allowed. They also controlled for the fundamental ability of both pitchers and hitters to isolate the actual "streakiness" of a player's performance.

A player's most recent twenty-five times at bat were a significant predictor of how that player would do at his next time up—good enough to justify an adaptive reaction by coaches. When a player is hot, the researchers calculated, his expected on-base percentage will be twenty-five to thirty points higher than it would be if he has been cold. Similarly, a player on a hot streak will be 30 percent more likely to hit a home run than if he has been on a cold streak.

References

Adams, Robert M. "The 'Hot Hand' Revisited: Successful Basketball Shooting as a Function of Intershot Interval." *Perceptual and Motor Skills* 74, no. 3 (June 1992): 934. http://journals.sagepub.com/doi/abs/10.2466/pms.1992.74.3.934.

Adams and Koehler Conley. 1992. "Investment Bankers Can Learn a Thing or Two from Athletes on a 'Hot Streak'" by Edmund Andrews. https://qz.com/193256/investment-bankers-can-learn-a-thing-or-two-from-athletes-on-a-hot-streak/.

Andrews, Edmund. "Investment Bankers Can Learn a Thing or Two from Athletes on a 'Hot Streak.'" *Quartz*, March 30, 2014. https://qz.com/193256/investment-bankers-can-learn-a-thing-or-two-from-athletes-on-a-hot-streak/.

Gilovich, Thomas, Robert Vallone, and Amos Tversky. "The Hot Hand in Basketball: On the Misperception of Random Sequences." *Cognitive Psychology* 17 (1985): 295–314. http://wexler.free.fr/library/files/gilovich%20(1985)%20the%20hot%20hand%20in%20basketball.%20on%20the%20misperception%20of%20random%20sequences.pdf.

Hale, Steven D. "An Epistemologist Looks at the Hot Hand in Sports." *Journal of the Philosophy of Sport* 26, no. 1 (1999): 79–87. Published online March 30, 2012. https://www.tandfonline.com/doi/abs/10.1080/00948705.1999.9714580.

Hooke, Robert. "Basketball, Baseball, and the Null Hypothesis." *Chance* 2, no. 4 (1989): 35–37. Published online September 20, 2012. https://www.tandfonline.com/doi/abs/10.1080/09332480.1989.10554952?journalCode=ucha20.

Koehler, Jonathan J., and Caryn Conley. "The 'Hot Hand' Myth in Professional Basketball." *Journal of Sport & Exercise Psychology* 25 (2003): 253–260. https://papers.ssrn.com/sol3/papers.cfm?abstract_id=1469609.

Larkey, Patrick D., Richard A. Smith, and Joseph B. Kadane. "It's Okay to Believe in the 'Hot Hand.'" *Chance* 2, no. 4 (1989): 22–30. Published online September 20, 2012. https://amstat.

tandfonline.com/doi/abs/10.1080/09332480.1989.10554950#. WseI0C7waCg.

Tversky, Amos, and Thomas Gilovich. "The Cold Facts about the 'Hot Hand' in Basketball." *Chance* 2, no. 1 (1989): 16–21. http://www.medicine.mcgill.ca/epidemiology/hanley/c323/ hothand.pdf.

Tversky, Amos, and Thomas Gilovich. "The 'Hot Hand': Statistical Reality or Cognitive Illusion?" *Chance* 2, no. 4 (1989): 31–34. Published online September 20, 2012. https://www.tandfonline. com/doi/abs/10.1080/09332480.1989.10554951.

Wardrop, Robert L. "Basketball." In *Statistics in Sport*, edited by Jay Bennett. London: Hodder Education Group, 1998.

Wardrop, Robert L. "Simpson's Paradox and the Hot Hand in Basketball." *American Statistician* 49, no. 1 (1995): 24–28. Published online February 27, 2012. https://amstat. tandfonline.com/doi/abs/10.1080/00031305.1995.10476107#. WseIcS7waCg.

Zwiebel, Jeffery, and Brett Green. "The Hot Hand Fallacy: Cognitive Mistakes or Equilibrium Adjustments? Evidence from Baseball." *Stanford Graduate School of Business*, November 2013. www.gsb.stanford.edu/faculty-research/working-papers/hot-hand-fallacy-cognitive-mistakes-or-equilibrium-adjustments.

Passing Matrix

Think about a basketball or an indoor soccer team: with five players in the field. Suppose that in a given situation (perhaps a set play), each player can pass the ball to only certain teammates. To represent this situation in matrix form, we could construct a five-by-five incidence matrix A, where the rows represent the passing player with the ball and columns represent the receiving player. We place a 1 in the ith row and jth column of this matrix if there is possibility of passing the ball to player j from player i; otherwise, insert a 0. We also place 0s on the principal diagonal, because player i cannot pass the ball to herself. With the situation represented in the form of a matrix, we can perform operations on this matrix to obtain information about the game plan.

$$
\begin{array}{c}
\text{Receiving Player} \\
\begin{array}{cccccc}
 & 1 & 2 & 3 & 4 & 5 \\
\end{array}
\end{array}
$$

$$
\text{Passing Player} \quad
\begin{array}{c}
1 \\ 2 \\ 3 \\ 4 \\ 5
\end{array}
\begin{bmatrix}
0 & 1 & 0 & 1 & 0 \\
0 & 0 & 1 & 0 & 0 \\
1 & 0 & 0 & 0 & 1 \\
0 & 0 & 1 & 0 & 0 \\
0 & 0 & 0 & 1 & 0
\end{bmatrix} = A
$$

(1) Find A^2. What does the 1 in row 2 and column 1 of A^2 indicate about the situation? What does the 2 in row 1 and column 3 indicate about the situation? In general, how would you interpret each element not on the principal diagonal of A^2?

(2) Find A^3. What does the 1 in row 4 and column 2 of A^3 indicate about the situation? What does the 2 in row 1 and column 5 indicate about the situation? In general, how would you interpret each element not on the principal diagonal of A^3?

(3) Compute A, $A + A^2$, $A + A^2 + A^3$,... , until you obtain a matrix with no 0 elements (except possibly on the principal diagonal), and interpret the result.

Solution

(1)

$$
A^2 =
\begin{bmatrix}
0 & 1 & 0 & 1 & 0 \\
0 & 0 & 1 & 0 & 0 \\
1 & 0 & 0 & 0 & 1 \\
0 & 0 & 1 & 0 & 0 \\
0 & 0 & 0 & 1 & 0
\end{bmatrix}
\begin{bmatrix}
0 & 1 & 0 & 1 & 0 \\
0 & 0 & 1 & 0 & 0 \\
1 & 0 & 0 & 0 & 1 \\
0 & 0 & 1 & 0 & 0 \\
0 & 0 & 0 & 1 & 0
\end{bmatrix}
=
\begin{bmatrix}
0 & 0 & 2 & 0 & 0 \\
1 & 0 & 0 & 0 & 1 \\
0 & 1 & 0 & 2 & 0 \\
1 & 0 & 0 & 0 & 1 \\
0 & 0 & 1 & 0 & 0
\end{bmatrix}
$$

The 1 in row 2, column 1 indicates that there is one way that player 2 can get the ball to player 1 with one intermediate pass–namely, 2 to 3 to 1. The 2 in row 1, column 3 indicates that there are two ways for player 1 to get the ball to player 3–namely, 1 to 2 to 3, and 1 to 4 to 3.

(2)

$$A^3 = AA^2 = \begin{bmatrix} 0 & 1 & 0 & 1 & 0 \\ 0 & 0 & 1 & 0 & 0 \\ 1 & 0 & 0 & 0 & 1 \\ 0 & 0 & 1 & 0 & 0 \\ 0 & 0 & 0 & 1 & 0 \end{bmatrix} \begin{bmatrix} 0 & 0 & 2 & 0 & 0 \\ 1 & 0 & 0 & 0 & 1 \\ 0 & 1 & 0 & 2 & 0 \\ 1 & 0 & 0 & 0 & 1 \\ 0 & 0 & 1 & 0 & 0 \end{bmatrix} = \begin{bmatrix} 2 & 0 & 0 & 0 & 2 \\ 0 & 1 & 0 & 2 & 0 \\ 0 & 0 & 3 & 0 & 0 \\ 0 & 1 & 0 & 2 & 0 \\ 1 & 0 & 0 & 0 & 1 \end{bmatrix}$$

The 1 in row 4, column 2 indicates that there is one way to get the ball from player 4 to player 2 with two intermediate passes. The 2 in row 1, column 5 indicates that there are two ways for player 1 to get the ball to player 5 with two intermediate passes.

(3) From parts (1) and (2)

$$A + A^2 = \begin{bmatrix} 0 & 1 & 0 & 1 & 0 \\ 0 & 0 & 1 & 0 & 0 \\ 1 & 0 & 0 & 0 & 1 \\ 0 & 0 & 1 & 0 & 0 \\ 0 & 0 & 0 & 1 & 0 \end{bmatrix} + \begin{bmatrix} 0 & 0 & 2 & 0 & 0 \\ 1 & 0 & 0 & 0 & 1 \\ 0 & 1 & 0 & 2 & 0 \\ 1 & 0 & 0 & 0 & 1 \\ 0 & 0 & 1 & 0 & 0 \end{bmatrix} = \begin{bmatrix} 0 & 1 & 2 & 1 & 0 \\ 1 & 0 & 1 & 0 & 1 \\ 1 & 1 & 0 & 2 & 1 \\ 1 & 0 & 1 & 0 & 1 \\ 0 & 0 & 1 & 1 & 0 \end{bmatrix}$$

$$A + A^2 + A^3 = \begin{bmatrix} 0 & 1 & 2 & 1 & 0 \\ 1 & 0 & 1 & 0 & 1 \\ 1 & 1 & 0 & 2 & 1 \\ 1 & 0 & 1 & 0 & 1 \\ 0 & 0 & 1 & 1 & 0 \end{bmatrix} + \begin{bmatrix} 2 & 0 & 0 & 0 & 2 \\ 0 & 1 & 0 & 2 & 0 \\ 0 & 0 & 3 & 0 & 0 \\ 0 & 1 & 0 & 2 & 0 \\ 1 & 0 & 0 & 0 & 1 \end{bmatrix} = \begin{bmatrix} 2 & 1 & 2 & 1 & 2 \\ 1 & 1 & 1 & 2 & 1 \\ 1 & 1 & 3 & 2 & 1 \\ 1 & 1 & 1 & 2 & 1 \\ 1 & 0 & 1 & 1 & 1 \end{bmatrix}$$

$$A^4 = AA^3 = \begin{bmatrix} 0 & 1 & 0 & 1 & 0 \\ 0 & 0 & 1 & 0 & 0 \\ 1 & 0 & 0 & 0 & 1 \\ 0 & 0 & 1 & 0 & 0 \\ 0 & 0 & 0 & 1 & 0 \end{bmatrix} \begin{bmatrix} 2 & 0 & 0 & 0 & 2 \\ 0 & 1 & 0 & 2 & 0 \\ 0 & 0 & 3 & 0 & 0 \\ 0 & 1 & 0 & 2 & 0 \\ 1 & 0 & 0 & 0 & 1 \end{bmatrix} = \begin{bmatrix} 0 & 2 & 0 & 4 & 0 \\ 0 & 0 & 3 & 0 & 0 \\ 3 & 0 & 0 & 0 & 3 \\ 0 & 0 & 3 & 0 & 0 \\ 0 & 1 & 0 & 2 & 0 \end{bmatrix}$$

$$A + A^2 + A^3 + A^4 = \begin{bmatrix} 2 & 1 & 2 & 1 & 2 \\ 1 & 1 & 1 & 2 & 1 \\ 1 & 1 & 3 & 2 & 1 \\ 1 & 1 & 1 & 2 & 1 \\ 1 & 0 & 1 & 1 & 1 \end{bmatrix} + \begin{bmatrix} 0 & 2 & 0 & 4 & 0 \\ 0 & 0 & 3 & 0 & 0 \\ 3 & 0 & 0 & 0 & 3 \\ 0 & 0 & 3 & 0 & 0 \\ 0 & 1 & 0 & 2 & 0 \end{bmatrix} = \begin{bmatrix} 2 & 3 & 2 & 5 & 2 \\ 1 & 1 & 4 & 2 & 1 \\ 4 & 1 & 3 & 2 & 4 \\ 1 & 1 & 4 & 2 & 1 \\ 1 & 1 & 1 & 3 & 1 \end{bmatrix}$$

This shows that it is possible to pass the ball from any player to any other player with at most three intermediate passes. In sum:

In A^2, the 1 in row 2, column 1 indicates that there is one way that player 2 can get the ball to player 1 with one intermediate pass—namely, 2 to 3 to 1. The 2 in row 1, column 3 indicates that there are two ways for player 1 to get the ball to player 3—namely, 1 to 2 to 3, and to 1 to 4 to 3.

In A^3, the 1 in row 4, column 2 indicates that there is one way to get the ball from player 4 to player 2 with two intermediate passes. The 2 in row 1, column 5 indicates that there are two ways for player 1 to get the ball to player 4 with two intermediate passes.

$A + A^2 + A^3 + A^4$ shows that it is possible to pass the ball from any player to any other player with at most three intermediate passes.

CHAPTER 2

Fun Facts

Turtles with a minimum physical activity live significantly longer than most active creatures.

THIS CHAPTER PRESENTS some interesting and somehow surprising occurrences in sports. It includes an article regarding exercise and one about fitness. It also looks at performances of some exceptional athletes.

Sports versus Regular Exercise

Many sports are used by active people as both exercise and social activity. These days, most people get their physical activities by going to gyms. Considering this, let me present a fun reading related to exercise and its benefits.

Recently, there has been some research on the possible association between physical activities and life expectancy for adults. To determine the number of years of life gained from physical activity, researchers have examined data on more than 650,000 adults. One study that appeared in *PLOS Medicine,* November 6, 2012, reports that people who engaged in physical activity had life expectancy gains of up to 4.5 years. Based on such studies, physical activities are recommended for anyone who can afford the time, energy, and cost.

While reading this and few similar studies, I was wondering about possible drawbacks of regular exercise, and I think I have found some. Of course, needless to say that this is not to encourage you to give up such activity. However, pointing them out may provide a few "comforting" arguments for those who, for any reason, do not or cannot exercise regularly.

Since so many different kinds of physical activities could be classified as exercise, to simplify the discussion here, I only consider more-serious ones that require traveling to a gym. Consider an individual who devotes ten hours of his/her weekly life to exercise (exercise plus travel to and from a gym). Suppose also that this individual does this through his/her adult life of, say, fifty years and with the goal of increasing his/her life expectancy. Simple calculation shows that for this person, time devoted to exercise would take almost three years of his/ her life. Moreover, since we cannot exercise while asleep, it takes, in fact, 4.5 years of our active day lives (say sixteen hours daytime). So the question that arises at this point is the following: Would exercise increase a person's life by more than three years already invested? The answer is it may or may not, and we cannot be sure about that. So here we are spending a sure period of our young life for an unsure gain of the end of life when we have less energy, fewer desires, and maybe less/lower health to enjoy what life has to offer. Of course, exercise could improve the quality of life, but for most people, regular exercise has its own physical, emotional, and mental stress. Just imagine spending three years of your life traveling to a gym or running on a treadmill. Other issues to consider are

- time taken from other activities;
- cost, especially for those who live in big cities;
- time to shower and to wash the gym clothes;
- possible injuries during the exercise;
- cost of buying gear such as shoes; etc.

Now, clearly, we can extend the discussion to many other related factors. For example, most of us enjoy casual recreations such as walking, biking, or swimming. These are probably the best form of physical activity that are helpful to both body and mind.

Here is another good news and interesting research finding. According to a recent research, drinking red wine is better than going to the gym. Jason Dyck and his colleagues in University of Alberta in Canada found that red wine, nuts, and grapes have a complex called resveratrol, which improves heart, muscle, and bone functions; the same way they're improved when one goes to the gym. Details of this research can be found in the article by Natalie Roterman | Sept 15 2014, 04:51PM EDT. www.latintimes. com › drinking-wine-better-going-gym.

So to those who have a busy life, maybe regular exercise is not always what we need to do, but as they say, we all need to get up, dress up, and show up, even if we do it as slow as a turtles that with a minimum physical activity live significantly longer than many apparently active creatures.

Who Is Normal? A Silly Joke

This section is about a silly joke. If you are familiar with Roman numerals, you probably know that XL = 40. You may have also heard that people who get to their forties use that as an excuse to be large size and overweight. Here is a story I have made up for being overweight. It is based on a theorem known as central limit theorem that plays the key role in science in general and in statistics in particular.

Central limit theorem is probably the most important and

celebrated theorem of statistics. It states that if you have a random sample of measurements from a population and add them up, the distribution of sum will tend to bell curve (normal distribution) as you increase the sample size.

I used the theorem to make up the following story. Thirty laborers and a foreman, who is in charge of hiring and firing, work in a remote farm. Every evening meals are delivered to the foreman by an outside contractor. The amount of calorie per package varies between 1,200 and 1,800. Before distributing, the foreman steals around 10 percent of each worker's share for his personal consumption.

Let X_i be the number of calories consumed by a worker. Then the number of calories consumed by the foreman is $Y = 0.10$ $(X_1 + X_2 + \ldots + X_{30})$. Having a hard job and eating only once a day, workers had a very flat belly, like the picture on the left. This is like uniform distribution. The question is what would be the shape of the foreman's belly? Since X_i's are independent random variables with a uniform distribution, according to the central limit theorem, distribution of Y is approximately normal as in the picture on the right. Foreman, of course, blames the theorem for his problem.

It is interesting to ask which individual we should name normal.

Who Is Fit?

One of the major questions asked by individuals and exercise physiologists are (1) how to define fitness and, more importantly, (2) how to measure it. It is well known that factors such as maximum heart rate and the time it takes for it to return to normal upon cessation of exercise are related to the person's physical condition. For example, the fitter the person is, the faster his or her pulse rate will return to normal after exercise. Since it is very easy to measure the pulse, many fitness tests have been developed based on recovery time after exercise.

The pulse ratio tests, as they are called, usually involves stepping up and down at a specified rate off a box of a specified height. One such test is the Harvard step test originally designed for use with male university students. It is a straightforward test to administer and has been adapted for groups other than male students. The values used for the height of the box and rate are dependent upon the age and sex of the person.

The test for adult males is as follows: the subject steps up and down off a box twenty inches high at a rate of thirty times per minute. When the subject steps up onto the box, he must attain a position in which the body is erect; crouching is not permitted. The stepping procedure involves four stages: left foot is placed on the box, the right foot is placed on the box, and then the left foot is placed on the floor, and the right foot is placed on the floor. The person is permitted to change the order of the feet provided that the order of the four stages and the rate of stepping are maintained.

The stepping continues for five minutes unless exhaustion is reached previously. Either case, the duration of stepping is recorded. Immediately after completion, the person sits down in a chair, and three pulse counts are taken (at the wrist) at following times: 1 to 1½, 2 to 2½, and 3 to 3½ minutes after stepping ceased. The person's Harvard index is then obtained from the formula

Harvard index = 100 x (duration of exercise in seconds) / (2 x sum of three pulse counts during recovery).

The physical condition, or fitness, of the subject is then determined according to the following scale. If the Harvard index is greater than 90, level of fitness is excellent; 80–89 is good; 55–79 is average; and less than 55 is poor. As an example, suppose that a Harvard step test using a highly trained marathon runner as a subject gave the following three pulse counts: 51, 48, and 43. The Harvard index is therefore

$$(300 \times 100) / 2(51 + 48 + 43) = 105.63.$$

so we conclude that the subject was extremely fit. Marathon runners can be expected to score high values of the Harvard index since the index is a measure of recovery from prolonged exercise, and the marathon is certainly a prolonged exercise.

The manager of sport teams often records their players' indices prior to and on completion of a period of training to see if training has improved the players' fitness.

In chapter 16, "Mathematics of Games and Competition, we wish to get a feel for how mathematics might be employed to help situation that involves conflict or competition. It is mostly about situations where one person's decision depends on other people's decisions. Think about games like chess or tennis, where your move is determined, at least in part, by what you think the other players could or would do. In a larger scale, the cold war was the stage for one of the original and most important applications of what is known in mathematics as the game theory. The basic idea of game theory is quite simple and mostly familiar.

Usain Bolt

Consider the records for men's one-hundred-meter run. For this event, twenty records are set since 1912. The last three records, 9.58, 9.69, and 9.72, were set in years 2009, 2008, and 2008, respectively—all by Bolt. Data for one-hundred- and two-hundred-meter runs exhibit long tail, as the present records for these events, 9.58 seconds and 19.19 seconds, respectively, are significantly lower than the previous records. Application to the men's one hundred meters data for the period starting January 1,

1977, when IAAF required fully automatic timing, to September 1, 2009, shows that the probabilities of setting a new record such as 9.55 seconds or less and 9.5 seconds or less are respectively

    A. 0.0102 and 0.0052, when Bolt's records are included;
    B. 0.0043 and 0.0023, when Bolt's records are excluded.

Also, excluding Bolt's records the probability of setting a record of 9.58 seconds or less by other runners is only 0.0064.

Application of the method to Bolt's individual performance prior to the 2008 Olympics reveals the following:

    A. For him, the probability of running the two hundred meters in the 19.30 seconds or less was only 0.00257, indicating that his Olympic record, 19.30 seconds, was completely unexpected.
    B. The probability of breaking his own best record, 19.75 seconds, was only 0.0738, indicating that his Olympic performance was exceptional.

Also, application of the method to his individual performance, including his 2008 Olympic record, reveals that his new record, 19.19 seconds, was even more astonishing.

Application of the method for estimation of ultimate record produces the following 90 percent prediction intervals:

    A. (9.40, 9.58) when Bolt's t records are included;
    B. (9.62, 9.71) when Bolt's records are excluded.

Note that Bolt's last record, 9.58, falls outside the interval B. This demonstrates that Bolt is in a different league.

Finally, the time between the last and penultimate records is one year, and the time the last record has held to date is zero. Excluding the last record (9.58), the probability of setting a record 9.58 or less was only 1.8 percent, which again indicates how extraordinary the new record is. Also, the probability of setting a record 9.55 or less was only 1.7 percent, and the probability of setting a record 9.50 or less was only 1 percent.

Michael Jordan

Data for basketball may be found in places such as the Pro Basketball Bible, nba.com, and espn.com. Data published includes number of games in which each player appeared, minutes per game (MPG), points per minute (PPM) played, field goal percentage (FGP), free throw percentage (FTP), assists per minute (APM), rebounds per minute (RPM), and so on. Suppose that performance is measured by points scored per minute played (PPM). Note that when comparing basketball players, we should consider the position they play, since their primary responsibility might be to distribute the ball rather than to score.

In *A Casebook for a First Course in Statistics and Data Analysis*, Chatterjee et al. analyzed data on 105 guards who played in the 1992–93 NBA season. Inspection of the data for guards showed that the distribution of PPM is close to a bell (normal) curve with mean and standard deviation of 0.4236 and 0.1159 points per minute, respectively. So using the table for normal distribution, the probability for a player to perform the same or better than Michael Jordan with PPM = 0.8291 for that season is 1/5,000 = 0.0002. This means that if fifty guards join NBA each year, it takes almost one hundred years to produce a player in that level, exceptional and impressive.

Mr. Inconsistent May Set a Record

A consistently inconsistent runner may give the coach what he wants.

Most coaches consider an athlete's consistency as a positive attribute, for obvious reasons. The performance of a consistent player exhibits less variability, and therefore such players are more predictable, something that coaches can utilize for their game plan. The following example illustrates how in some sports, variability could prove helpful for a different reason.

Think about the top two members of a college track team, A and B. Suppose that the coach has kept their statistics and has concluded that their times, in seconds, follow a bell curve (normal distribution) with means of 10.4 and 10.6 and standard deviation of 0.2 and 0.4, respectively (N [10.4, 0.2] and N [10.6, 0.4]). Suppose also that the college competition's best/present record is ten seconds. The coach thinks that on a good day, both runners have a chance to break the record, but he's not sure which runner has the better chance. His assistant thinks that it is A, who has averaged 10.4, which is closer to 10.

The coach is wondering about another factor—namely, B's inconsistency, because he has heard that athletes with larger mean value and greater irregular performance (larger standard deviation) can have a higher chance to drop below a chosen threshold. So he uses the tables for normal distribution and calculates the probability of breaking the record for A and B. He finds that they are 0.0228 and 0.0668, respectively, which shows that the athlete with larger mean value and greater irregular performances (larger standard deviation) have a better chance of breaking the ten-second record.

To better appreciate the difference, suppose that both runners can participate in ten tournaments during a season. Based on the above calculations, the probabilities that for the season, neither of them would run under ten seconds are respectively $0.794 = (1 - 0.0228)^{10}$ and $0.501 = (1 - 0.0668)^{10}$. This implies that their probabilities of breaking the record in one season are respectively 0.206 and 0.499.

We can analyze this situation differently by considering distribution of the minima of samples of sizes to ten from a standard normal or any other parent population. This can be done through what is known as order statistics, which is the data ordered in ascending order.

It's interesting that the athlete with larger mean time and frequent irregular performance (larger standard deviation) can have a better chance to run below ten seconds. In other words, the latter athlete, though more irregular and with smaller mean time, should be chosen for the competition if only one athlete must be chosen. This example illustrates how extreme values look and feel different from central value problems.

Comparing Athletes

People who follow sports compare their favorite athletes. This is not easy, especially if these athletes are from different sports or different years. However, this may be done using relative standing of the athletes by comparing them to other players in the same sport.

Z-Score

Suppose that x presents a performance measure in a specific sport. For a sample of size n, $\{x_1, x_2, ..., x_N\}$ of such measure, we can calculate the sample mean x_{avg} and the sample standard deviation s. Then for each measurement x_i, the sample z-score is calculated by subtracting the mean and dividing it by standard deviation—that is, $z_i = (x_i - x_{avg})/s$. By definition, we see that the z-score of a value tells how many standard deviations that value is away from (either above or below) the mean. Since standard deviation is positive, it follows that if the z-score is positive, the value is above the mean; if the z-score is negative, the value is below the mean; and if the z-score is equal to zero, then the value is exactly equal to the mean. Thus, z-score provides some information regarding the relative standing of a given measurement. Note that if we change the unit of measurements by adding or multiplying the data values by a fixed number, the face values of x will change but not their z-scores. In other words, z-scores are scale-invariant. The mean of the z-scores is always 0, and the standard deviation of the z-scores is always 1.

Example

Suppose we wish to compare two students taking the same course at the same university, one in the morning and the other in the afternoon class, to determine who the better student is. Suppose that the following information is available.

Description	Grade	Class Average	Class Standard Deviation	z-score
Student in the morning class	80	85	5	$z = (80 - 85)/5 = -1$
Student in the afternoon class	76	72	4	$z = (76 - 72)/4 = 1$

From this table, it is clear that, although the first student has a higher grade, she has a lower z-score, which means that the second student has a much better relative standing. In other words, the grade of the second student is one standard deviation above the average grade in her class; whereas the first student's grade is one standard deviation below the average grade in her class. This means that the second student has done better than many more students in her class than the first student in her class.

Since we don't know what the grades represent, a reasonable approach is to compare students using their z-score, which indicates their relative standing. This eliminates factors such as hard tests, bad teachers, and poor textbooks. In other words, each student is judged based on his or her relative standing in their own class. We can apply the same procedure when comparing athletes from different sports or students from different majors or universities.

I Want to Be Like Mike

As mentioned earlier, data for basketball may be found in places such as the Pro Basketball Bible, nba.com, and espn.com. In *A Casebook for a First Course in Statistics and Data Analysis*, Chatterjee et al. analyzed data on 105 guards who played in the 1992–93 NBA season. Inspection of the data for guards showed that the distribution of PPM is close to a bell (normal) curve with mean and standard deviation of 0.4236 and 0.1159 points per minute, respectively. So we can calculate the z-score for Michael Jordan, whose PPM for that season was 0.8291, as

$z = (0.8291 - 0.04236)/0.1159 = 3.5$.

Plotting the data reveals that the distribution of PPM is approximately normal. Using normal distribution table, we find that approximately 99.7 percent of guards had PPM to within three standard deviations of the mean. So a z-score of 3.5 is extremely unusual and quite impressive.

Comparison of Baseball Players

In the 1910s, 1940s, and 1970s, batting averages were 0.266, 0.267, and 0.261, respectively, with standard deviations of 0.037, 0.0326, and 0.0317. Suppose we want to compare three players and decide who should be ranked highest. Obviously, we have to compare like with like, so the best way to rank them would be in relationship to their contemporaries. The players' names and batting averages are Ty Cobb (0.420 in 1911), Ted Williams (0.4064 in 1941), and George Brett (0.390 in 1980). We calculate their z-scores as 4.151, 4.264, and 4.07, respectively. So Ted Williams was ranked as the best hitter, since he has a higher z-score—that is, a better relative standing.

Comparison of Soccer Teams

As a different example, consider the English premiership data for the year 2001. The mean and standard deviation of the points (three points for a win, one point for a tie, and zero for a loss) earned by twenty teams were respectively fifty-two and fourteen. The z-score for Manchester United, with eighty points, was 2. The nearest team, Arsenal, had a z-score of 1.29. This shows the strength of Manchester United in that season and provides a reason for their popularity.

References

Barry, Rick. 1995. *Rick Barry's Pro Basketball Bible, 1995–96: Player Ratings and In-Depth Analysis of More Than 400 NBA Players and Draft Picks.* Marina Del Ray, CA: Basketball Books.

Chatterjee, Samprit, Mark S. Handcock, and Jeffrey S. Simonoff. 1995. *A Casebook for a First Course in Statistics and Data Analysis.* Hoboken, NJ: Wiley.

NBA Advanced Stats. https://stats.nba.com/.

"NBA Statistics." ESPN. http://www.espn.com/nba/statistics.

CHAPTER 3

Tennis and Table Tennis

Among the popular sports, tennis and table tennis lend themselves to mathematical analysis better than most other popular sports.

Tennis History

RETIRED BRITISH MAJOR Walter Wingfield invented *lawn tennis* in 1873. He borrowed heavily from *court tennis*, a French game that had been played by the British aristocracy for centuries, and *badminton*, a game that had more recently come to Britain from India. Court tennis was played in special indoor courts. Major Wingfield's primary innovation was to move tennis outside, *a la* badminton. Those who said, "I could have invented this game," criticized him. Still his "tennis sets" (balls, four rackets, and a net) sold well, and lawn tennis caught on quickly in Britain. The first "world tennis championship" was held at Wimbledon in 1877, and soon the game spread to the United States and throughout the British Empire. Today, tennis is popular worldwide, and Wimbledon is truly a world championship. Perhaps one reason tennis is so popular is players of all ages and abilities can enjoy the game. Even weekend players can experience the satisfaction of serving an ace, "putting away" a volley at the net, or watching a well-placed lob sail over the head of a helpless opponent.

Tennis has also become popular with mathematicians. Calculating the odds of winning a tennis match is a fun exercise in mathematical modeling. The British mathematician Ian Stewart wrote an amusing dialog along these lines called "The Drunken Tennis-Player" for *Pour la Science*, the French version of *Scientific American*. Stewart's article was later reprinted in his book *Game, Set and Math*. Another nice discussion of mathematical modeling of tennis appears in *Mathematics and Sports* by Russian mathematicians L. E. Sadovskii and A. L. Sadovskii.

Stewart's "drunken tennis-player" makes some calculations on a napkin in the pub after a game and exclaims, "The rules of tennis favor the better player." Is this true? As his opponent responds, "But they should, shouldn't they? I mean, the better player ought to have a better chance of winning?" A better way to phrase the question we will examine is "Do the quirks of the rules of tennis inflate the stronger player's chance of winning?" We will attempt to answer the question by building several mathematical models of a hypothetical tennis match between two players: Reza and Bill. By changing the rules of tennis in some of our models, we can investigate how rule changes would affect the odds of winning. Millionaire tennis player, tennis organizer, and tennis innovator James Van Allen promoted many of the alternative scoring systems we discuss.

In all of our models, we will assume that Reza has a 60 percent probability of winning any given point. We assume that the 60 percent probability remains constant throughout the match. This assumption is clearly an oversimplification, particularly since it is likely that Reza's probability of winning a point is higher when he is serving than when he is receiving. However, as we will discuss later, these models behave in essentially the same way as more complicated models that allow the probability of winning a point to vary from point to point. The advantage of using a simple model is that it makes it easy to calculate probabilities. Most of the calculations needed for the results in this paper can be done on a hand calculator. We used the computer software Mathematica, since it makes it easy to do even the more complicated calculations.

The Quirks of Scoring

Major Wingfield proposed that lawn tennis use badminton scoring: matches of fifteen points with players scoring only when serving. This innovation was largely ignored. During the early years, tennis players used a variety of scoring systems. By the time of the first championship at Wimbledon, the All England Croquet Club had settled on a scoring system based on the traditions of court tennis. This system remained unchanged until the introduction of tiebreakers in 1970.

One quirk of tennis scoring is that strange names are used for points in scoring a game: *love, fifteen, thirty, forty,* and *game.* Although no one knows the origin of this odd system, it has been proposed that *fifteen, thirty, forty-five, sixty* were originally used to represent the four quarters of an hour. Over the years, the score *forty-five* became abbreviated as *forty.* (In informal play, *fifteen* is sometimes abbreviated as *five.*) It would be simpler to score the game: *zero, one, two, three, and four.* Still, the weird point names give no advantage to either player.

A more important quirk is that two points must win a game. If Reza and Bill have each scored three points, the score is called *deuce,* rather than *40:40.* If Reza wins the next point, the score becomes *advantage Reza.* If Reza wins again, he wins the game; otherwise, the score returns to *deuce.* This feature of tennis scoring does increase the chance that the stronger player will win, as we shall see.

Game versus No-Ad Game

To see how the "win by two rule" affects the probability of the stronger player winning a game, we will construct two simple mathematical models. First, we will model an alternate version of a tennis game, proposed by Van Allen, called a *no-ad game*. The winner of a no-ad game is the first player to win four points. It is not necessary to win by two. For comparison, we will also model a standard tennis game.

What is Reza's probability of winning a no-ad game? Figure 1 shows all scores that are possible in a no-ad game.

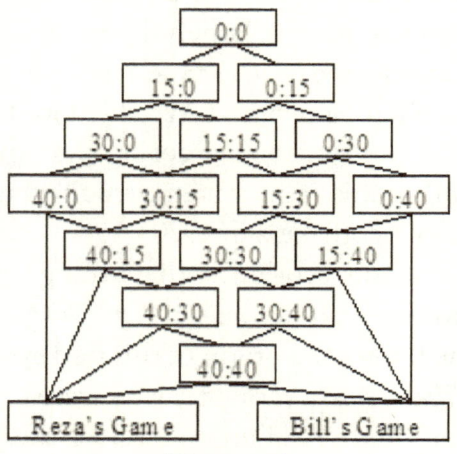

Figure 1

We always list Reza's score first. For example, 15:30 means Reza has scored one point and Bill has scored two points. As each point is played, we descend one level of the chart, with a 60 percent probability of moving to the left and a 40 percent probability of moving to the right. After the first point, the probability is 60 percent that the score is 15:love and 40 percent that the score is love:15. After the second point, there are three possible scores: 30:love, 15:15, and love:30. The probability that Reza will win both of the first two points is the product of the probabilities of winning each point. We will write P(30:love) for the probability that the score will be 30:love after two points. In this notation, P(30:love) = 0.6 · 0.6 = 36%. Similarly, P(love:30) = 0.4 · 0.4 = 16%. There are *two ways* of reaching a score of 15:15–Reza scores first and Bill scores second, or vice versa. So P(15:15) = 0.6 · 0.4 + 0.4 · 0.6 = 48%. The following table summarizes our results for the three possible scores after two points.

Score	*30:love*	*15:15*	*love:30*
Probability	*36%*	*48%*	*16%*

By continuing to calculate in this way, we may find the probability of each possible score in a no-ad game and, ultimately, the probability that Reza will win. We find that

$$P \text{ (Reza wins a no} - \text{ad game)} = 0.6^4(1 + 4 \cdot 0.4 + 10 \cdot 0.4^2 + 20 \cdot 0.4^3) = 71\%$$

The numbers 1,4,10, and 20 come from the fact that, while there is only one way to reach the score 40:love, there are four ways to reach 40:15, ten ways to reach 40:30, and twenty ways to reach 40:40. Notice that the probability that Reza wins a no-ad game is *higher* than his 60 percent probability of winning a point.

Deuce Problem

How can we model the "win by two" rule used in the standard game? We start by replacing 40:40 in figure 1 by deuce, as shown in figure 2.

Replace

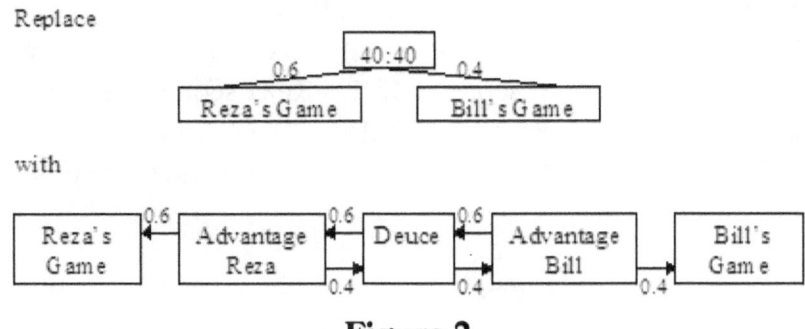

with

Figure 2

Now, there is no limit to how long the game could go on, alternating between *advantage Reza, deuce,* and *advantage Bill.* In principle, the game could go on *forever,* but the probability of that is zero. How can we calculate infinitely many probabilities? Fortunately, we do not have to—algebra comes to the rescue. The infinitely many probabilities we need to calculate form a *geometric series,* an infinite sum with a constant ratio between the terms. A little algebra allows us to find the sum of our geometric series. Let $x = $ P (Reza eventually wins, starting from deuce). If Reza wins the next two points, he will win the game. If Bill and Reza split the next two points, the score will be deuce again, and Reza's probability of winning the game will be x again. It follows that x satisfies the equation.

$x = $ P (Reza wins the next two points) + P (Reza and Bill split the next two points) x

This may seem to be a circular definition of x, but it allows solving for x algebraically. We have

$$x = 0.6 \cdot 0.6 + (0.6 \cdot 0.4 + 0.4 \cdot 0.6)x$$

Algebraically simplifying the equation, we find that $0.52x = 0.36$. Hence $x = 32/56 = 69.2\%$.

We can now simplify the picture of deuce as shown in figure 3.

Replace

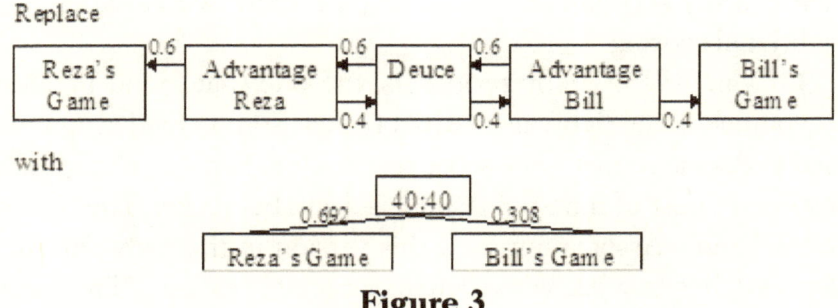

with

Figure 3

Notice that the "win by two rule" favors Reza, since it increases P(Reza wins, starting from 40:40) from 60 percent to 69.2 percent. However, the effect of the "win by two rule" does not appear as large when one looks at the probabilities of winning a game. Reza's probability of winning a standard game turns out to be 73.6 percent, while his probability of winning a no-ad game is 71 percent, a difference of only 2.6 percent. The "win by two" rule *does* help Reza, but only if the score reaches 40:40. Dropping the rule would only change the outcome of 2.6 percent of games between Bill and Reza.

Tradition versus Predictability

Traditionally, the first player to win six games and lead by two games wins a *set*. After 5:5, the set goes on until one player gets ahead by two, sometimes for a very long time. The longest match in professional tennis history took place at Wimbledon in 2010. It lasted eleven hours and five minutes (played over the course of three days), easily eclipsing the previous record of six hours and three minutes set during the 2004 French Open. The Wimbledon match was decided after 980 points, and, most amazingly, the final set (which does not end until one of the players obtains a two-game lead) reached a score of 70–68. The improbability of this score motivates our investigation of win-by-two games. To avoid such lengthy sets, James Van Allen promoted two variations on traditional scoring.

Of Van Allen's many proposals, the one that found greatest acceptance is the tiebreaker. A tiebreaker is an extended "game" used to decide the set if the score reaches 6:6. Van Allen originally proposed a best-of-nine-point sudden-death tiebreaker. The United States Tennis Association used this version in the early '70s, but later switched to what Van Allen disparagingly called a "lingering-death" tiebreaker; the first player to win seven points *and lead by two points* wins. The acceptance of the tiebreaker was the first change in the official rules of tennis scoring in almost a century. Tradition bowed to the need to make the length of televised tennis matches more predictable.

Suppose that Bill and Reza are tied at 6:6. Would Bill have a better chance of winning if they play a tiebreaker or if they follow the traditional rules? In the traditional game, the score of 6:6 is analogous to *deuce*, as we see in figure 4.

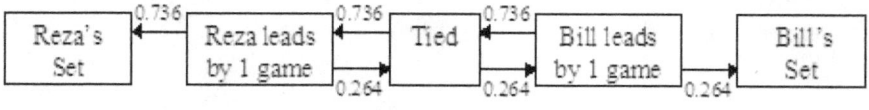

Figure 4

The only difference is that we must replace the probability of winning a point by the probability of winning a game. We can find x=P (Reza wins, starting from 6:6) in the same way we found P (Reza wins starting from deuce). We solve the equation $x = 0.736^2 + 2 \cdot 0.736 \cdot 0.264x$ and find that $x = 88.7\%$.

What happens if Bill and Reza use a tiebreaker? We may calculate P(Reza wins the tiebreaker) in the same way we found P(Reza wins a game). The result is that P(Reza wins the tiebreaker) = 78.7%. Thus, the tiebreaker gives Bill a somewhat better chance of winning. However, if we now use these results to determine the probability of winning a set, we find that Reza's probability of winning a traditional set is 96.6 percent, while his probability of winning a set with a tiebreaker is 96.3 percent. Using a tiebreaker will change the outcome of only about 0.3 percent of the sets between Reza and Bill.

The Van Allen Streamlined Scoring System

A more radical variation on the rules for scoring sets is the Van Allen Streamlined Scoring System (VASSS), devised by Van Allen in 1958. To simplify the scoring and speed up matches, Van Allen scrapped games entirely. The first player to score thirty-one points wins the set, with a best-of-nine sudden-death tiebreaker at 30:30. The serve alternates every five points as in table tennis. To calculate the probabilities of winning a VASSS set, we use the same kind of calculations we used for the no-ad game. The result is that Reza has a 94.7 percent probability of winning a VASSS set. Although VASSS is not likely to please tennis traditionalists, it does not usually change the outcome of a set. Only about 2 percent of sets between Bill and Reza will have different outcomes under VASSS than under traditional scoring.

Beyond VASSS

Traditionally, there are two kinds of tennis matches: a three-set match and a five-set match. The winner is the first to score two sets in a three-set match or three sets in a five-set match. If scrapping games doesn't make much difference, why not *really* streamline tennis scoring by scrapping sets as well? A match could simply be scored in points. Play would continue until one player reaches a certain number of points, say 150. We will call such a match a 150-match. The graph below produced using Mathematica shows the probability of winning a five-set match (using traditional scoring) as a function of the probability of winning a point.

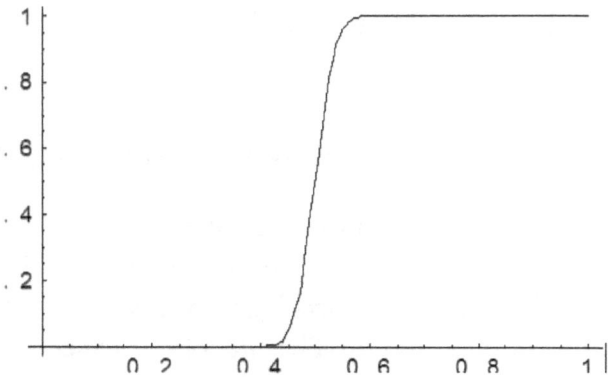

Since Reza has a 60 percent probability of winning a point, his probability of winning the match is 99.96 percent! The steepness of the graph seems to support the thesis that the rules of tennis inflate the stronger player's chance of winning. Is there really something special about the rules of tennis that is working in the stronger player's favor? In the examples we have examined so far, Reza's strong advantage has been robust to a number of variations in the scoring. In the next graph, we plot the probability of winning a 150-match along with the probability of winning a traditional five-set match.

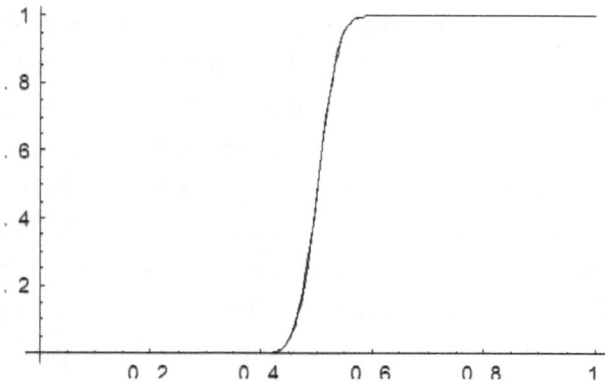

The results are almost identical, indicating that it is not really the rules of tennis that gives the stronger player such a high probability of winning. Reza's probability of winning a 150-match is 99.98 percent.

The Law of Large Numbers

What is actually guaranteeing the stronger player's high probability of victory is a fundamental result of statistics known as the *law of large numbers*. The law of large numbers predicts that the longer the players play, the more likely it is that the stronger player will win. This is the same law that predicts that a gambler who keeps betting against odds that slightly favor the house will ultimately be ruined.

More precisely, the law of large numbers says that *whenever* a random phenomenon is repeated a large number of times, the proportion of times each outcome occurs approaches the probability of that outcome. If Reza and Bill play a large number of points, we can expect Reza to win close to 60 percent of the points. As the number of points increases, the probability approaches one that Reza will win nearly 60 percent of the points.

Thus, in any kind of match in which Bill and Reza play a large number of points, Reza will have a high probability of ending with a significantly higher number of points than Bill. Let us define a *reasonable* scoring system to be one that rarely allows a player to win if he is behind in points. Then *any* reasonable scoring system will award Reza the victory almost all the time. Therefore, it is not the quirks of tennis scoring that "favor the stronger player." The rules need only ensure that a match contains many points. The law of large numbers does the rest!

Should We Believe the Models?

The predictions of any mathematical model must be viewed with some skepticism. A mathematical model simplifies reality and thereby distorts the truth. A good mathematical model is one that is simple enough to use to make predictions but close enough to reality that the predictions of the model are useful. How well do our models reflect what actually happens in a tennis match? One glaring omission is that we have failed to take the effect of the serve into account. Although having to serve is sometimes a negative factor for a beginner, experienced players have a higher probability of winning the games they serve. Usually, the player who is serving will win the game. Sets are close because the players alternate serving. The turning point in a set is often a service break, where the receiving player wins a game. Thus, the alternation of the serve would appear to somewhat reverse the advantage of the stronger player and give the weaker player a better chance. As the "drunken-tennis player" says,

> The way the scoring amplifies any advantage means that each player has a chance rather close to one of winning his service game–provided his chance of winning a point is above 1/2. That tends to act the *opposite* way, which evens the game out again!

It is not terribly hard to redo our analysis of a set, considering changing probabilities of winning a point as the player's alternate serving. For instance, we can look at a model where Reza has an 80 percent probability of winning a point when he is serving, but only a 40 percent probability when he is receiving. In this model, Bill has a 60 percent probability of winning a point on his serve and a 73.6 percent probability of winning his service games. This model predicts that Reza's probability of winning the match is 96.2 percent, a little less than in the model in which we ignored the serve. However, the difference is small enough that it confirms the conclusions we made using the simpler model. Since the serve alternates, the 80 percent and 40 percent balance out in the end, giving us results consistent with a fixed probability of 60 percent. Below are two graphs plotted on the same axes. For one of the graphs, the horizontal axis represents Reza's fixed probability, p, of scoring a point. The vertical axis represents his probability of winning a set. The second graph shows the result of considering the serve. In this model, Reza's probability of scoring a point is $p + 0.2$ when he is serving and $p - 0.2$ when he is receiving. The fact that the graphs are almost identical confirms that we can safely ignore the serve.

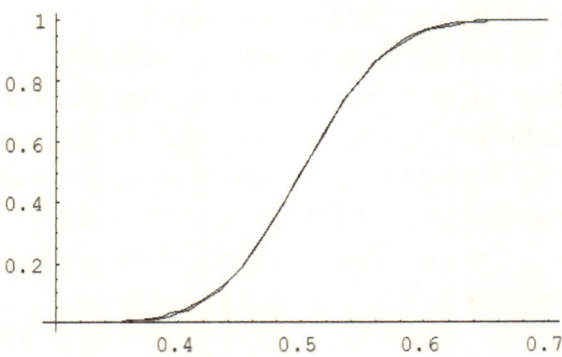

Perhaps a more serious problem with the model is that the results do not seem to be consistent with experience. According to the model, Bill will essentially never be able to beat Reza. Even if Bill could improve his probability of winning a point from 40 percent to 45 percent, the model predicts that he would only win one out of every twenty-five matches. According to the model, the usual situation is for one player to dominate the other. However, it is not uncommon for a player who loses a match to win a rematch. According to our model, this will only happen if the players are extremely well matched. It seems unlikely that the players are always so well matched.

A better explanation of the fact that different players win on different occasions is that the probability of winning a point does not stay fixed each time the players play but fluctuates as the players' level of practice or concentration varies from day to day. For instance, suppose Bill and Reza play every day of the week. Bill's probability of winning a point is 55 percent on Monday, but

37.5 percent the other days of the week. Then, on average, Bill will win $(55\% + 6 * 37.5\%)/7 = 40\%$ as before. (For simplicity, we are assuming that each match contains the same number of points.) However, Bill's winning points are now not evenly distributed. He has a high probability of winning the Monday match and losing the others. Thus, he would win about $1/7^{th}$ of the matches instead of less than $1/1000^{th}$ of them.

Still, even if the probability of winning a point fluctuates from match to match, it is not unreasonable to suppose that the probability of winning a point is constant throughout a given match. Even when we incorporated the fluctuation of the probabilities from game to game, as the serve alternates, we saw very little change in the probabilities of winning. Therefore, we have a good deal of confidence that our comparisons of different scoring systems are valid.

Conclusions

The answer to the question of whether the "drunken tennis-player" is correct in saying that the "rules of tennis favor the better player" is yes and no. Yes, in that the rules call for the players to play a large number of points in a match. No, in that any reasonable scoring system with a large number of points will produce the same results.

What can we say about changing the rules? Our models predict that the use of a tiebreaker changes the outcome of a set very rarely. The tiebreaker helps keep matches from running too long. It is no wonder that it has been widely adopted. We could go further and replace games with no-ad games or sets with VASSS sets or even matches with 150-matches without altering the outcome of the match in most cases. Doing so would simplify the scoring and probably speed up play, but there is something to be said for tradition. Deuce and tiebreakers do make tennis more exciting to play and to watch, even if they do not often change the outcome of the match. Traditional scores are also easier to remember. We imagine players of a 150-match would often have to stop to ask questions such as "Is it 127:109 or 126:109?"

Finally, some comments about winners and losers. Our models show that it is unlikely for the weaker player to win a tennis match purely by chance. Thus, if Reza wins the match, he is very likely the better player, at least for that day. However, Bill can take consolation in the fact that the loser of a tennis match may be only slightly weaker than his opponent. Even though Reza has beaten Bill consistently in the past, Bill only needs to win one more point in ten to be even with Reza. With a little practice, Bill may be able to win the next match!

Tennis and Calculus

Let us discuss an interesting way of teaching derivative using tennis. In the above calculation, Reza had a $20\% = (60\% - 40\%)$ edge in one point, which increased to $47.2\% = (73.6\% - 26.4\%)$ edge in a standard game. For good players, the edge is usually very small, may be only a percent or two. If we call the edge in a point Δx, the edge in a game (Δy) can be calculated using derivatives. This provides an efficient and very interesting way of teaching derivatives including the chain rule by considering sets and matches. Suppose that Reza and Bill have arrived at deuce. Let P denote the probability that Reza wins that game. Recall that this could happen if Reza takes the next two points with probability x^2 or they split the first two points with probability $x(1-x) + (1-x)x = 2x(1-x)$ and return to deuce. Thus,

$$P = x^2 + 2x(1-x)p$$

or

$$P = \frac{x^2}{1 - 2x(1-x)} = \frac{x^2}{x^2 + (1-x)^2}$$

Using this example for $x = 0.55$, for example, we get $P = 0.60$.

This shows that a 10% = (55% - 45%) advantage (edge) in a point offer arriving at deuce open up to 20% (60% - 40%) in a game. In a similar way, we can calculate edge in a set and a match. This can be used to teach chain rule for calculating derivatives of a function of a function.

References

Collins, Bud, and Zander Hollander (editors). 1994. *Bud Collins' Modern Encyclopedia of Tennis*. Detroit: Gale Research Inc.

Robertson, Max, and Jack Kramer (editors). 1974. *Encyclopedia of Tennis*. New York: Viking Press.

Sadovskii, L. E., and A. L. Sadovskii. 1993. *Mathematics and Sports*, translated from the Russian by S. Markar-Liamanov. Providence: American Mathematical Society.

Stewart, Ian. 1991. *Game Set and Math*. London: Penguin Books.

United States Lawn Tennis Association. 1972. *Official Encyclopedia of Tennis*. New York: Harper and Row.

Table Tennis

As mentioned earlier most sports can be analyzed using mathematics, statistics, and computer sciences. Many can be utilized to enhance teaching of mathematical topics and concepts. In fact, sports and the data related to them offer a unique opportunity to teach mathematics and statistics and to test the methods they offer. This section presents a probabilistic analysis of a table tennis game with a view toward its possible use as an aid for demonstrating steps of mathematical modeling. The materials presented are suitable for further analysis and study. We begin by describing the history and the rules of the game.

History and Rules

Table tennis, also known as ping-pong, has its origins dating back to medieval tennis. Beginning as a mere social diversion, table tennis became popular in England during the latter part of the nineteenth century. It did not take long for the popularity of this game to spread. As early as 1901, table tennis tournaments were being organized, books about table tennis were written, and even local table tennis associations were formed. In 1902, an unofficial "World Championship" was held. Table tennis was widely popular in Central Europe from 1905 to 1910. A slightly altered version was introduced prior to this to Japan, which then spread to China and Korea.

Today, table tennis has become a major worldwide sport, with approximately thirty million competitive players and countless others who enjoy playing the sport at a leisurely level. Presently, the International Table Tennis Foundation represents 140 countries. In 1988, it became an Olympic sport.

Table tennis is an object sport in which each competitor tries to control an object while the other competitor is in direct confrontation. The object of the game is to hit the ball into the opponent's table. If the ball is not returned after one bounce, a point is won. The only other way to win a point is when the opposing player commits an error by striking the net or by hitting the ball off the table. The table must be 2.74 meters (9 feet) long and 1.525 meters (5 feet) wide. The top of the net must be 15.25 centimeters (6 inches) above the playing surface, while the posts that secure the net are 15.25 centimeters beyond the side line. The spherical ball must weigh 2.7 grams (0.09524 ounces) with a diameter of 40 millimeters (1.574 inches). The ball may be white or orange, in either case not glossy, and it can be made of celluloid or similar plastic material. While most rackets seem to be the same shape, it can actually be of any size, shape, or weight as long as the blade is flat and rigid. Natural wood must comprise 85 percent of the blade by thickness. Neither can be thicker than 7.5 percent of the total thickness or 0.35 millimeters, whichever one is smaller. A game is won by the player who either reaches eleven points by a margin of two or gains a lead of two points after both players have scored ten points. This scoring scheme was implemented in 2001. A match is simply the best of any odd number of games. A new serving system was also put into place in 2001. Each player serves for two points and then alternates with the other player until the end of the game. If the game reaches ten all, then the number of consecutive serves for each player is reduced to one. Concerning the serve, another new rule was put into effect in 2002. This rule states that the ball must be visible at all times–meaning, the server cannot block the receiver's view of the ball in any way (e.g. body, clothing, and table).

Figure 1 displays possible outcomes (states) of a game of table tennis under the new rules. Prior to 2001, each game consisted of twenty-one points instead of eleven and five serves instead of two.

Figure 1. Table tennis scoring illustration (new rules)

```
                              0-0
                          1-0    0-1
                      2-0    1-1    0-2
                  3-0    2-1    1-2    0-3
              4-0    3-1    2-2    1-3    0-4
          5-0    4-1    3-2    2-3    1-4    0-5
      6-0    5-1    4-2    3-3    2-4    1-5    0-6
  7-0    6-1    5-2    4-3    3-4    2-5    1-6    0-7
8-0  7-1    6-2    5-3    4-4    3-5    2-6    1-7    0-8
9-0  8-1    7-2    6-3    5-4    4-5    3-6    2-7    1-8    0-9
10-0  9-1   8-2    7-3    6-4    5-5    4-6    3-7    2-8    1-9    0-10
11-0  10-1  9-2    8-3    7-4    6-5    5-6    4-7    3-8    2-9    1-10   0-11
11-1  10-2  9-3    8-4    7-5    6-6    5-7    4-8    3-9    2-10   1-11
11-2  10-3  9-4    8-5    7-6    6-7    5-8    4-9    3-10   2-11
11-3  10-4  9-5    8-6    7-7    6-8    5-9    4-10   3-11
11-4  10-5  9-6    8-7    7-8    6-9    5-10   4-11
11-5  10-6  9-7    8-8    7-9    6-10   5-11
11-6  10-7  9-8    8-9    7-10   6-11
11-7  10-8  9-9    8-10   7-11
11-8  10-9  9-10   8-11
11-9  10-10  9-11
A's game   Advantage A   Tie   Advantage B   B's game
```

Probability of Winning a Game

Table tennis can be related to many mathematical topics. We start exploring these relationships by first focusing on some probability questions.

Suppose that a game is played by two players, A and B. Let the probability of A winning the point be denoted by x, and the probability of B winning the point by y, where $x + y = 1$, and both are independent of the present score. In practice, x and y may be estimated from players' past games against each other.

Before proceeding further, we need to comment on question of possible advantage associated with serving and its effect on probability of winning or losing a game. The assumption of fixed probability of winning a point is a standard one and is usually made for simplicity. However, it is not a serious one and its effect on probability of winning or losing a game is not as pronounced as it is for tennis since the server has to hit her own side of the table first. In fact, as is pointed out in an article titled "How Long Is an 11-Point Game?", djmarcusetasc.com (Wed. Feb. 10 10-39-01 1993) in most cases, the assumption of fixed probability has no effect on conclusion. Also, in a more detailed study titled "Does It Matter Who Serves First?", djmarcusetasc.com (Wed. Feb. 10 10-39-02 1993) which appeared in page 31 of Jan/Feb 91 *TTTopics*, the answer to the question posed in the title is stated as no based on the following argument. If the game goes tie (deuce), then it does not matter who served first, since no matter who wins, each player will have served the same number of times. For other cases, consider the following modifications of the rules. Rather than stopping when one player reaches eleven, keep playing until twenty points have been played. If player A wins the game under modified rules, then she must win at least eleven of the twenty points and hence would have won the game under standard rules. Similarly, if A loses under the modified rules, she also would have lost under the standard rules. But under the modified rules, both players serve ten times, and so it doesn't matter which one served first.

In the rest of this article, we will assume that x is fixed. However, we will return to this question later and make further comments before closing. For simplicity, we will only calculate the probability of A winning a game. Following the new rules of the game, A can either win by reaching a score of eleven with a margin of two (case 1) or, if players tie at 10:10, by winning extra points to have a margin of two points (case 2). Thus, the probability of A winning a game is the sum of these two probabilities—that is,

$$P(A \text{ winning a game}) = P(\text{Case 1}) + P(\text{Case 2}).$$

Let us discuss each case in more detail.

Case 1: In this case, A can win the game by any of the following scores 11:0, 11:1, 11:2, 11:3, 11:4, 11:5, 11:6, 11:7, 11:8, or 11:9. For these, A must win ten out of the respective 10, 11, . . . 19 points played, plus the last point. Using binomial distribution, it is easy to see that the probability of A winning with the score of 11:j is $\binom{10+j}{10}x^{11}y^j$, $j = 0, 1, ..., 9$. Thus, the probability of A winning the game by reaching a score of eleven with a margin of two is

$$P(\text{Case 1}) = \sum_{j=0}^{9} \binom{10+j}{10} x^{11} y^j = \sum_{j=0}^{9} \binom{10+j}{10} x^{11}(1-x)^j$$

Case 2: In this case, A and B must first reach the score of 10:10. Then A must gain a lead of two points to win the game. The probability of this event is therefore

$$P(\text{Case 2}) = P(A \text{ and } B \text{ both reach } 10) \times P(A \text{ wins by a margin of } 2).$$

First, we calculate the first term in the right-hand side. Since to reach 10:10, a total of twenty points must be played, we have

$$P(A \text{ and } B \text{ both reach } 10) = \binom{20}{10} x^{10} y^{10}$$

Next, to win the game, A must gain a lead of two points. This can be done in infinitely many ways until A finally gains a lead of two. Let p denote the probability that A wins the game after reaching the score of 10:10. Then we have

$$p = x^2 + 2xyp$$

This is because starting from 10:10, A can either takes the next two points with the probability x^2, or each player take a point with probability of $2xy$ and restart from the score of 11:11. The probability of A winning a game restarting from 11:11 is the same as starting from the score of 10:10. Solving for p, we get

$$p = \frac{x^2}{1 = 2xy}$$

Putting these together, the probability that A wins a game after reaching a score of 10:10 is

$$x = 0.5 - 0.9$$

$$P(\textbf{Case 2}) = \binom{20}{10} x^{10} y^{10} \frac{x^2}{1 - 2xy}$$

Replacing $1 - x$ for y, and $x^2 + (1 - x)^2$ for $1 - 2xy$, the probability of case 2 is

$$P(\textbf{Case 2}) = \binom{20}{10} \frac{x^{12}(1 - x)^{10}}{(x^2 + (1 - x)^2)} = \frac{184756(1 - x)^{10} x^{12}}{1 - 2x + 2x^2}$$

Finally, $g(x)$ the probability of A winning a game is obtained by adding the probabilities for the two cases—that is,

$$g(x) = \textbf{P(A wins)} = \textbf{P(Case 1)} + \textbf{P(Case 2)}$$

$$= \sum_{j=0}^{9} \binom{10+j}{10} x^{11}(1-x)^{j} + \binom{20}{10} x^{12}(1-x)^{10}/(x^2 + (1-x)^2)$$

Under the old rules, the probability of A winning a game denoted by $g_0(x)$ is

$$g_0(x) = \sum_{j=0}^{19} \binom{20+j}{20} x^{21}(1-x)^{j} + \binom{40}{20} x^{22}(1-x)^{20}/(x^2 + (1-x)^2)$$

Table 1 lists values of P(case 1) and P(case 2) and the total $(g(x))$ for different values of x for the new rules. Table 2 lists the same values for a game played according to the old rules, where, rather than eleven, players had to score twenty-one points to win. Table 3 and table 4 present different probabilities under the new and the old rules, respectively.

Table 1. Values of P(case 1), P(case 2) and the total probabilities $(g(x))$ for $x = 0.5 - 0.9$ (new rules).

s	0.5	0.505	0.51	0.525	0.55	0.6	0.7	0.8	0.9
11 to 0	0.0005	0.0005	0.0006	0.0008	0.0014	0.0036	0.0198	0.0859	0.3138
11 to 1	0.0027	0.003	0.0033	0.0044	0.0069	0.016	0.0653	0.189	0.3452
11 to 2	0.0081	0.0088	0.0096	0.0124	0.0186	0.0383	0.1175	0.2268	0.2071
11 to 3	0.0175	0.0189	0.0204	0.0256	0.0363	0.0664	0.1527	0.1965	0.0897
11 to 4	0.0305	0.0327	0.0350	0.0426	0.0572	0.0930	0.1603	0.1376	0.0314
11 to 5	0.0458	0.0486	0.0515	0.0606	0.0772	0.1116	0.1443	0.0825	0.0094
11 to 6	0.0611	0.0642	0.0673	0.0768	0.0926	0.1190	0.1154	0.0440	0.0025
11 to 7	0.0742	0.0771	0.0801	0.0886	0.1012	0.1156	0.0841	0.0214	0.0006
11 to 8	0.0835	0.0859	0.0883	0.0947	0.1025	0.1040	0.0568	0.0096	0.0001
11 to 9	0.0881	0.0898	0.0913	0.0950	0.0974	0.0879	0.0360	0.0041	3E-05
Case 1	0.4119	0.4296	0.4475	0.5015	0.5914	0.7553	0.952	0.9974	1
Case 2	0.0881	0.0898	0.0913	0.0945	0.0955	0.0811	0.026	0.0019	6E-06
Total	0.5	0.5194	0.5387	0.596	0.6868	0.8364	0.9781	0.9993	1

Table 2. Values of P(case 1), P(case 2) and the total probabilities $(g(x))$ for $x = 0.5 - 0.9$ (old rules).

Score	0.500	0.505	0.510	0.525	0.550	0.600	0.700	0.800	0.900
21-0	0.000	0.000	0.000	0.000	0.000	0.000	0.001	0.009	0.109
21-1	0.000	0.000	0.000	0.000	0.000	0.000	0.004	0.039	0.230
21-2	0.000	0.000	0.000	0.000	0.000	0.001	0.012	0.085	0.253
21-3	0.000	0.000	0.000	0.000	0.001	0.002	0.027	0.131	0.194
21-4	0.000	0.000	0.000	0.001	0.002	0.006	0.048	0.157	0.116
21-5	0.001	0.001	0.001	0.002	0.003	0.012	0.072	0.157	0.058
21-6	0.002	0.002	0.002	0.004	0.007	0.021	0.094	0.136	0.025
21-7	0.003	0.004	0.004	0.006	0.012	0.032	0.108	0.105	0.010
21-8	0.006	0.007	0.007	0.011	0.018	0.045	0.114	0.073	0.003
21-9	0.009	0.011	0.012	0.016	0.027	0.058	0.110	0.047	0.001
21-10	0.014	0.016	0.017	0.023	0.036	0.069	0.099	0.028	0.000
21-11	0.020	0.022	0.024	0.031	0.046	0.078	0.084	0.016	0.000
21-12	0.026	0.029	0.031	0.040	0.055	0.083	0.067	0.009	0.000
21-13	0.033	0.036	0.039	0.048	0.063	0.084	0.051	0.004	0.000
21-14	0.041	0.043	0.046	0.055	0.069	0.082	0.037	0.002	0.000
21-15	0.047	0.050	0.053	0.061	0.072	0.077	0.026	0.001	0.000
21-16	0.053	0.056	0.058	0.065	0.073	0.069	0.018	0.000	0.000
21-17	0.058	0.060	0.062	0.067	0.071	0.060	0.011	0.000	0.000
21-18	0.061	0.063	0.064	0.068	0.068	0.051	0.007	0.000	0.000
21-19	0.063	0.064	0.065	0.066	0.063	0.042	0.004	0.000	0.000
Case 1	0.437	0.462	0.488	0.564	0.684	0.870	0.0994	1.000	1.000
Case 2	0.063	0.064	0.065	0.066	0.061	0.038	0.003	0.000	0.000
Total	0.500	0.526	0.553	0.629	0.746	0.909	0.997	1.000	1.000

Table 3. Different probabilities under the new rules.

x	Tie	A wins after tie	A wins with tie	A wins without tie
0.000	0.000	0.000	0.000	0.000
0.050	0.000	0.003	0.000	0.000
0.100	0.000	0.012	0.000	0.000
0.150	0.000	0.030	0.000	0.000
0.200	0.002	0.059	0.000	0.001
0.250	0.010	0.100	0.001	0.004
0.300	0.031	0.155	0.005	0.017
0.350	0.069	0.225	0.015	0.053
0.400	0.117	0.308	0.036	0.128
0.450	0.159	0.401	0.064	0.249
0.500	0.176	0.500	0.088	0.412
0.550	0.159	0.599	0.095	0.591
0.600	0.117	0.692	0.081	0.755
0.650	0.069	0.775	0.053	0.878
0.700	0.031	0.845	0.026	0.952
0.750	0.010	0.900	0.009	0.986
0.800	0.002	0.941	0.002	0.997
0.850	0.000	0.970	0.000	1.000
0.900	0.000	0.988	0.000	1.000
0.950	0.000	0.997	0.000	1.000
1.000	0.000	1.000	0.000	1.000

From these tables, it is clear that, under the old rules, the game will virtually never reach a tie unless x (or y) is between 0.35 and 0.65. If players are equally good, the probability of reaching a tie takes the maximum value of 0.125.

Similarly, under the new rules, the game will virtually never reach a tie unless the probability that a player wins a point is between 0.25 and 0.75. This means that a tie is more likely to occur under the new rules than under the old rules. If players are equally good, the probability of reaching a tie takes the maximum value of 0.176. Note that, although reaching a tie is more likely under the new rules, it is clear that the probabilities of winning the game after reaching a tie are the same under both rules.

Figure 2 presents the graphs of $g(x)$ and $g_0(x)$. As can be seen, $g(x) \geq g_0(x)$ for $0 < x \leq 0.5$ and $g(x) < g_0(x)$ for $0.5 < x \leq 1$. This means that the probability of winning the game is higher for the weaker player under the new rules, making the game less predictable and more exciting.

Table 4. Different probabilities under the old rules.

x	Tie	A wins after tie	A wins with tie	A wins without tie
0.000	0.000	0.000	0.000	0.000
0.050	0.000	0.003	0.000	0.000
0.100	0.000	0.012	0.000	0.000
0.150	0.000	0.030	0.000	0.000
0.200	0.000	0.059	0.000	0.000
0.250	0.000	0.100	0.000	0.000
0.300	0.004	0.155	0.001	0.002
0.350	0.019	0.225	0.004	0.017
0.400	0.055	0.308	0.017	0.074
0.450	0.103	0.401	0.041	0.213
0.500	0.125	0.500	0.063	0.437
0.550	0.103	0.599	0.061	0.684
0.600	0.055	0.692	0.038	0.870
0.650	0.019	0.775	0.015	0.964
0.700	0.004	0.845	0.003	0.994
0.750	0.000	0.900	0.000	0.999
0.800	0.000	0.941	0.000	1.000
0.850	0.000	0.970	0.000	1.000
0.900	0.000	0.988	0.000	1.000
0.950	0.000	0.997	0.000	1.000
1.000	0.000	1.000	0.000	1.000

Figure 2. Graphs of $g(x)$ (dashed curve) and (solid curve).

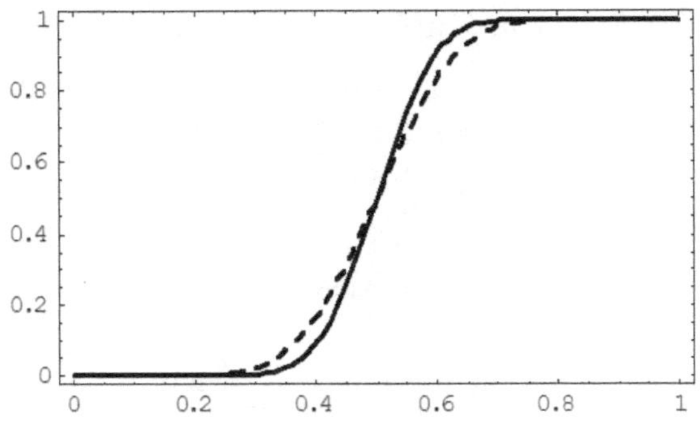

Analysis Using Difference Equations

Let us once more consider the event that A and B reach score of 10:10. We can use difference equations to find the probability that A wins the game starting from this score or, in fact, any other possible score. Here, we use the tie situation for demonstration.

Suppose that $g(i,j)$ represents the probability of A winning the game starting from a score of $i{:}j$–that is, A has won i points and B has won j points. Since either player can win the next point, this implies that

$$g(i,j) = xg(i + 1,j) + yg(i,j + 1)$$

Now, consider the case of a tie–that is, the score of 10:10. Using this relationship, we get

$$g(10,10) = xg(11,10) + yg(10,11),$$
$$g(11,10) = xg(12,10) + yg(11,11),$$

and

$$g(10,11) = xg(11,11) + yg(10,12).$$

Substitution for $g(11,10)$ and $g(10,11)$ in the first equations yields

$$g(10,10) = x[xg(12,10) + yg(11,11)] + y[xg(11,11) + yg(10,12)]$$
$$= x^2g(12,10) + 2xyg(11,11) + y^2g(10,12)$$

If the game reaches a score of 12:10, then A has won the game. Therefore, $g(12,10) = 1$. Also, if the game reaches a score of 10:12, then B has won the game. This means that $g(10,12) = 0$. Finally, if the game reaches a score of 11:11, the situation is no different than the score of 10:10. Therefore, we have $g(11,11) = g(10,10)$. Replacing for these, we obtain

$$g(10,10) = x^2 + 2xyg(10,10)$$

or

$$g(10,10) = x^2/(1 - 2xy)$$

The value of $g(10,10)$ represents the probability that A wins the game starting from a score of 10:10. Note that this is the same as what we obtained in the previous section. Following the same lines, we can obtain the probabilities for other starting scores. For example,

$$g(11,10) = xg(12,10) + yg(11,11)$$
$$= x + yx^2/(1 - 2xy) = x(1 - xy)/(1 - 2xy)$$
$$g(10,11) = xg(11,11) + yg(10,12) = xg(10,10) = x^3/(1 - 2xy)$$
$$g(9,9) = g(10,10) = x^2/(1 - 2xy)$$
$$g(10,9) = g(11,10), \qquad g(9,10) = g(10,11)$$

Markov Chains

A Markov chain is a mathematical system that experiences transitions from one state to another according to certain probabilistic rules. The defining characteristic of a Markov chain is that no matter how the process arrived at its present state, the possible future states are fixed. In other words, the probability of transitioning to any particular state is dependent solely on the current state and time elapsed. For example, think of a tennis game. Suppose that in a game currently played, the score is 30–30. The probability of moving to the state of 40–30 does not depend on how the game has arrived in the state of 30–30. The state space, or set of all possible states, can be anything: letters, numbers, weather conditions, tennis scores, or stock performances.

Markov chains may be modeled by finite state machines, and random walks provide a prolific example of their usefulness in mathematics. They arise broadly in statistical and information-theoretical contexts and are widely employed in economics, game theory, queueing (communication) theory, genetics, and finance. While it is possible to discuss Markov chains with any size of state space, the initial theory and most applications are focused on cases with a finite number of states.

Transition Matrix

As pointed out, Markov chain is used to describe the "change" of state of a system as time passes. For example, suppose that John, who loves outdoor sports, keeps track of the weather in the area he lives. For simplicity, he only considers two states—namely "sunny" and "rainy." He is planning to play tennis tomorrow, but today the weather is in state "rainy." By keeping the records, John is trying to figure out the chance that tomorrow's weather would be in state "sunny." Here, John is dealing with a system that has only two states, and changes of states may happen from one day to the next. As another example, the result of the latest game played by two teams must be one of three possible states: state 1, team A won the game; state 2, tie; or state 3, team B won the game. In a playoff series or when considering games from one season to the next, it is possible that team A wins the first game and then team B wins the second one. Such a situation happens in sports like tennis or volleyball when several games or sets are played one after another or in basketball, ice hockey, etc., during the playoffs. In such situations, we may observe a "change of state"–that is, as "time" passes and we move from the first game to the second game, the state may change from state 1 into state 3. Note that in some sports, there is no tie, and therefore we need to consider only two states. In general, from the past records, it is possible to estimate the conditional probability of moving to one state given that in the previous period "time," we were in that or in another state. For example,

P(Tomorrow is Sunny | Today is Rainy)
P(Lisa will be in a good mood tomorrow
| She is in a good mood today)
P(Having "hot hands" in tomorrow's game |
Had a "hot hand" in today's game),

or

P(Team B wins the second game ?| Team A wins the first one).

The latter is exactly the probability for the state to change from 1 into 3 and is conventionally denoted by p_{13}. Similarly, all the probabilities can be estimated from the past records. Here, p_{ij} denotes the probability that the state changes from i to j, in unit of time or period, etc.

In general, if a process can be in one of the k possible states, labeled $1, 2, \ldots, k$ and if the probability p_{ij} that the state changes into state j at any time after it was in state i at the preceding time can be determined, then the process is called a Markov chain. The probability p_{ij} is called transition probability from state i to state j. The matrix $P = [p_{ij}]$ is called the transition matrix of the Markov chain. Note that we can also think of Markov chain as a sequence of stochastic moves such that the probability for the next state is completely determined by the present state.

For example, suppose that John has a record of the last ten days as

SSRSRRRSSR.

Then, since S(sunny) followed S twice, R(rainy) followed S

three times, S followed R twice, and R followed R twice, we have the following (estimate) for the transition matrix:

$$\begin{array}{cc} & \begin{array}{cc} S & R \end{array} \\ \begin{array}{c} S \\ R \end{array} & \begin{bmatrix} 0.4 & 0.6 \\ 0.5 & 0.5 \end{bmatrix} \end{array}$$

A practice problem

Many people believe in "hot hands" in sport. This thinking gives rise to the belief that streaks, or "hot hands," are unusual. In fact, streaks are to be expected in random sequences, and long streaks needn't be amazing coincidences. As an example, here is the 1999 win-loss record for the Baltimore Orioles:

WLLLWLLLWLLLLLLWLLLWLLWLWWWWWLWLLLLLLLWWLLWLWLWLWWL
WLLWLLWLLLWWWWWWWWLWWWWLLLLLLLLLLLWWLLLLLWWWWW
LLWWWWWWLWLLLLLiLWWLWLLWWLLLWWWLLWWWLLLLLWLW
WLWLLWWWWWWWWWWWWLWWLW

The team's overall average was almost exactly 0.500. Use this to find a transition matrix. Can you draw any conclusion?

Steady state solution

We start by noting that if all elements of the transition matrix or its powers are positive, the corresponding Markov chain is called a regular Markov chain.

If a Markov chain is regular, then the process approaches to a fixed state vector q. The vector q is called the steady state vector of the regular Markov chain. Note that this vector does not require the knowledge of an initial state vector. In other words, it is the same for any arbitrary initial state vector.

Absorbing states

In this section, we use two examples from sports to present

the concept of the absorbing Markov chain. The first example involves tennis.

In a tennis game played by players A and B, after the state "deuce" is reached, the game becomes a Markov chain involving five states: A's game, B's game, deuce, advantage A, and advantage B. Let us number these state as 1, 2, 3, 4, and 5, respectively. If we further assume that the probability of winning a point by either player is fixed and equals to x and $y = 1 - x$ –that is,

$$x = P(\text{player A wins the point})$$

And

$$y = 1 - x = P(\text{player B wins the point})$$

then we have the following transition matrix:

$$P = \begin{array}{c} \\ 1 \\ 2 \\ 3 \\ 4 \\ 5 \end{array} \begin{array}{ccccc} 1 & 2 & 3 & 4 & 5 \\ \begin{bmatrix} 1 & 0 & 0 & 0 & 0 \\ 0 & 1 & 0 & 0 & 0 \\ 0 & 0 & 0 & x & y \\ x & 0 & y & 0 & 0 \\ 0 & y & x & 0 & 0 \end{bmatrix} \end{array}$$

Note that in this example, both p_{11} and p_{22} equal 1, reflecting the fact that once we reach these states (A's game or B's game), everything is over and the process will never change its state again. Thus, once in such a state, the process stays in that state forever.

Further analysis

In section 3, we calculated the probability for A to win a game as a function of x. In this section, we study certain aspects of this function. Recall that

$$g(x) = \sum_{j=0}^{9} \binom{10+j}{10} x^{11}(1-x)^j + \binom{20}{10} x^{12}(1-x)^{10} / (x^2 + (1-x)^2)$$

Let us examine the graph of g(x) over the interval [0,1] shown in figure 2.

It is clear that g(x) is an increasing function over the interval [0,1]. This means that as x increases, the probability of A winning the game also increases. The minimum occurs at x = 0, where the probability of A winning a point is 0. The maximum occurs at g(x) = 1, where the probability of A winning a point is 1. The first derivative of g(x) is

$$g'(x) = 369512(1-x)^{10} x^{10} \left(5-9x+9x^2\right) / \left(1-2x+2x^2\right)^2$$

Figure 3 depicts a representation of this function and $g_0'(x)$. Values of g(x) for few x's are shown in table 6.

Figure 3. Graph of g'(x) (dashed curve) and $g_0'(x)$ (solid curve).

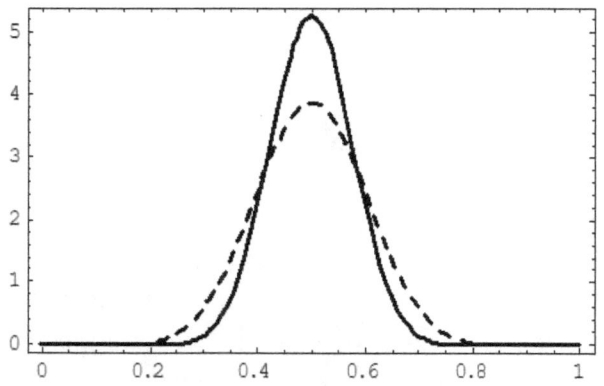

Table 6. Selected values of g(x).

x	g'(x)	x	g'(x)
0.5000	3.8763	0.8000	0.0313
0.6000	2.4607	0.9000	0.0001
0.7000	0.5698	1.0000	0.0000

Since g'(x) > 0 for all $0 \le x \le 1$, g(x) is increasing on this interval. This indicates that as x increases, the probability of A winning a game also increases. The rate of increase for the probability of A winning a game is maximum at x = 0.5, where g'(x) equals 3.876. This indicates that around x = 0 for every 1 percent change in x, g(x) changes by 3.876 percent. Thus, whenever one player is slightly better than her opponent, her chance of winning a game changes a great deal. So around x = 0.50, a small advantage for either player will result in a significant advantage in a game. On the other hand, for x values close to 0 or 1, the rate of change for her chance of winning will not change a great deal because g'(x) has low values at these extremes. This means that for these values, the chance of A winning a game is already very high, or very low, and thus any small change in x will not affect A's chance of winning significantly. Finally, since g'(x) decreases over the interval [0.5,1], it follows that the rate of change for the probability of A winning a game is decreasing.

We also have

$$g''(x) = \frac{-369512 \; x^9 (1-x)^9 \left(-50+279x-699x^2+1006x^3-810x^4+324x^5\right)}{\left(1-2x+2x^2\right)^3}$$

Figure 4. Graph of g''(x) (dashed curve) and $g_0''(x)$ (solid curve).

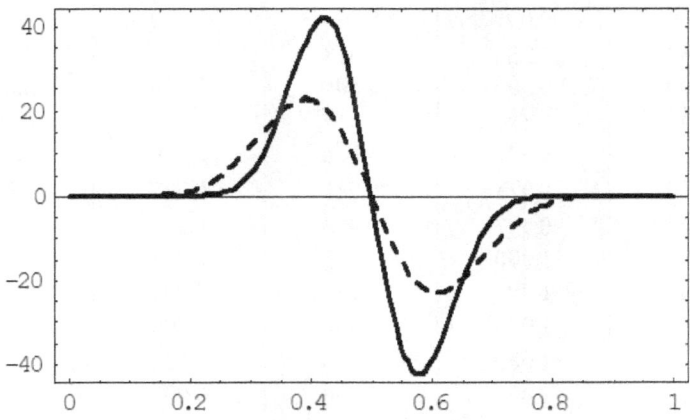

from which it can be determined that there is an inflection point at x = 0.50. The graph of this function and $g_0''(x)$ are given in figure 4. When x < 0.5, g''(x) > 0, indicating that g(x) is concave up on (0, 0.5). Since g'(x) > 0, this means that the rate of increase for the probability of A winning the game increases on this interval. However, when x > 0.5, g''(x) < 0, meaning that g(x) is concave down on (0.5, 1). This indicates that while A's chances of winning the game is increasing on this interval, the rate of increase decreases as it approaches its maximum of 1.

The information provided so far can be used to compare the probabilities and rate of changes under the new and old rules. Table 7 provides a summary. As expected, the weaker player has a better chance under the new rules.

Table 7. Probabilities and rates of changes under the new and old rules.

x	$g_O(x)$	$g_N(x)$	$g_O'(x)$	$g_N'(x)$
0.000	0.000	0.000	0.000	0.000
0.050	0.000	0.000	0.000	0.000
0.100	0.000	0.000	0.000	0.000
0.150	0.000	0.000	0.000	0.003
0.200	0.000	0.001	0.001	0.031
0.250	0.000	0.005	0.013	0.168
0.300	0.003	0.022	0.137	0.570
0.350	0.022	0.069	0.727	1.364
0.400	0.091	0.164	2.230	2.461
0.450	0.254	0.313	4.260	3.465
0.500	0.500	0.500	5.266	3.875
0.550	0.746	0.687	4.260	3.465
0.600	0.909	0.836	2.230	2.461
0.650	0.978	0.931	0.727	1.364
0.700	0.997	0.978	0.137	0.570
0.750	1.000	0.995	0.013	0.168
0.800	1.000	0.999	0.001	0.031
0.850	1.000	1.000	0.000	0.000
0.900	1.000	1.000	0.000	0.000
0.950	1.000	1.000	0.000	0.000

We end this section by noting that the analysis presented may help in teaching some mathematical concepts. For example, calculation of g?(x) can be related to the idea of one player having a slight advantage (edge) over the other player in one point and her edge in one game. The instructor can ask the students to calculate g?(x) for values close to x = 0.5 by calculating the difference between the g(x) values and also by using the definition directly.

Analysis Using Markov Chain

In this section, we use matrices and a Markov chain to calculate the probability that A wins a game after reaching the score of 10:10. The analysis presented here addresses the first passage characteristics of the Markov chain. First, we set up a transition matrix with elements representing the probabilities of moving from a state to other states. Let state 1 indicate that A has won the game, and state 2 indicate that B has won the game. Let state 3 refer to tie. Let state 4 indicate that A has a point advantage over B, and state 5 indicate that B has a one-point advantage over A.

State 1	State 4	State 3	State 5	State 2
A's Game	Advantage A	Tie	Advantage B	B's Game

The transition matrix denoted by T is then

$$T = \begin{array}{c} \\ 1 \\ 2 \\ 3 \\ 4 \\ 5 \end{array} \begin{array}{ccccc} 1 & 2 & 3 & 4 & 5 \\ \left(\begin{array}{cc|ccc} 1 & 0 & 0 & 0 & 0 \\ 0 & 1 & 0 & 0 & 0 \\ \hline 0 & 0 & 0 & x & y \\ x & 0 & y & 0 & 0 \\ 0 & y & x & 0 & 0 \end{array}\right) \end{array}$$

Note here that states 1 and 2 are absorbing states, whereas the other three states are transient.

Suppose now that the game is in state 3. We can represent this by a state vector taking the form

$$(0 \ \ 0 \ \ 1 \ \ 0 \ \ 0)$$

Multiplication by transition matrix T yields

$$(0 \ \ 0 \ \ 0 \ \ x \ \ y)$$

This is the state vector after playing one point. It is easy to show that the repeated multiplication leads to a final state vector of the form

$$(x^2/(1-2xy) \quad y^2/(1-2xy) \ \ 0 \ \ 0 \ \ 0) = (x^2/(x^2+y^2) \quad y^2/(x^2+y^2) \ \ 0 \ \ 0 \ \ 0)$$

This means that, eventually, one of the players will win the game. However, depending on the value of x, the time it takes for this to happen will vary. In fact, it is possible to determine the average length of the time it takes for each player to win starting from one of the three non-absorbing states (3, 4, and 5). For this, we divide the transition matrix into four matrices, I, 0, R, and Q–that is,

$$T = \begin{pmatrix} I & 0 \\ R & Q \end{pmatrix}$$

where I and 0 are identity and zero matrices, respectively. Here, I, 0, R and Q are matrices of dimensions (2x2), (2x3), (3x2) and (3x3), respectively. Next, we find the matrix $F = (I - Q)^{-1}$ known as the fundamental matrix. This matrix determines how long it would take, on average, to reach an absorbing state starting from a nonabsorbing state. Here, we have

$$I - Q = \begin{pmatrix} 1 & -x & -y \\ -y & 1 & 0 \\ -x & 0 & 1 \end{pmatrix}$$

Its inverse $F = (I - Q)^{-1}$ is then

$$
\begin{array}{c}
3 \\
4 \\
5
\end{array}
\begin{pmatrix}
\dfrac{1}{1-2xy} & \dfrac{x}{1-2xy} & \dfrac{y}{1-2xy} \\[2mm]
\dfrac{y}{1-2xy} & \dfrac{1-xy}{1-2xy} & \dfrac{y^2}{1-2xy} \\[2mm]
\dfrac{x}{1-2xy} & \dfrac{x^2}{1-2xy} & \dfrac{1-xy}{1-2xy}
\end{pmatrix}
$$

This indicates, for example, that starting from state 3 (a tie score), it takes an average of $= (1 + x + y)/(xy)$ points for the game to end. The sum of the other two rows provides similar information for games starting from states 4 and 5.

Finally, we calculate the matrix FR. This matrix provides probability of A or B winning the game starting from a nonabsorbing state. It is

$$
FR = \begin{array}{c}
3 \\
4 \\
5
\end{array}
\begin{array}{cc}
\quad 1 \qquad\qquad 2 \\
\begin{pmatrix}
\dfrac{x^2}{1-2xy} & \dfrac{y^2}{1-2xy} \\[2mm]
\dfrac{x(1-xy)}{1-2xy} & \dfrac{y^3}{1-2xy} \\[2mm]
\dfrac{x^3}{1-2xy} & \dfrac{y(1-xy)}{1-2xy}
\end{pmatrix}
\end{array}
$$

For example, when the score is tied at 10:10, the probability that A wins the game equals $x^2/(1-2xy)$. This is exactly what we obtained using the other approaches. Also, the probability that B wins starting from state 3 is $y^2/(1-2xy)$. Other entries are the corresponding probabilities for the games starting from states 4 or 5. Table 5 presents numerical values of some of these quantities for different values of x.

Table 5. Conditional probability of A winning and expected number of points given a tie or advantage situation.

x	Probability that A wins starting at state 3at state 3	Probability that A wins starting at state 4at state 4	Expected number of points before game ends starting at state 3	Expected number of points before game ends starting at state 4
0.00	0.000	0.000	2.000	3.000
0.05	0.053	0.000	2.210	3.099
0.10	0.111	0.001	2.439	3.195
0.15	0.176	0.005	2.685	3.282
0.20	0.247	0.012	2.941	3.353
0.25	0.325	0.025	3.200	3.400
0.30	0.409	0.047	3.448	3.414
0.35	0.496	0.079	3.670	3.385
0.40	0.585	0.123	3.846	3.308
0.45	0.671	0.180	3.960	3.178
0.50	0.750	0.250	4.000	3.000
0.55	0.820	0.329	3.960	2.782
0.60	0.877	0.415	3.846	2.538
0.65	0.921	0.504	3.670	2.284
0.70	0.953	0.591	3.448	2.034
0.75	0.975	0.675	3.200	1.800
0.80	0.988	0.753	2.941	1.588
0.85	0.995	0.824	2.685	1 .403
0.90	0.999	0.889	2.439	1.244
0.95	1.000	0.947	2.210	1.110
1.00	1.000	1.000	2.000	1.000

Additional Studies

The analysis of the game can be expanded in many different directions. This section includes few suggestions for this.

A. It is possible to determine the size of a handicap to make a game a fair game. For this, we first ask the following question. How many points should be accorded the weaker player for the game to be fair? Here, it is possible to construct a table to determine the number of handicap points that will nearly equalize player's chances of winning using a variable such as x / y = r.

We can also consider the inverse problem. Again, we ask the following questions; if B has been given a handicap of *j* points to make the game fair, what does that mean for the relative strength of the players? Here, we need to solve the equation $g(0, j) = \frac{1}{2}$ with respect to $r = x / y$ for a given value of *j*. This leads to an algebraic equation in *r*.

As pointed out, in practice, it is also possible to make the game fair by moving the net away from the middle of the table. To find out how much net should be moved, one can try different amounts and let players play games until several successive games end with close scores.

B. In the above analysis, we assumed that the probability of winning a point is fixed for each player throughout the match. Suppose that the probability of winning the point is not constant during the match. For example, a player may have a higher probability of winning the point when serving. If a game is played under a set of conditions that remains unchanged during the game, then it is possible to carry the analysis using an equivalent game with a fixed probability of winning a point. Here, equivalent means that the two games have the same chance of being won or lost. To clarify, consider the possible advantage associated with serving. Suppose that the probability for player A to win a point is p when serving and is q when receiving. We can describe this situation using the following transition matrix.

$$\begin{bmatrix} p & 1-p \\ q & 1-q \end{bmatrix}$$

Here, the fixed probability of winning a point, x, in the equivalent game may be taken as the stationary solution of a Markov chain having above transition matrix—that is, $q/(1 - p + q)$. If players are equally good servers—that is, $p = 1 - q$—then $x = 0.50$. Also, if $p > 1 - q$, then $x > 0.50$. For example, for $p = 0.60$ and $q = 0.45$, we get $x = 0.529$. This means that if A and B play according to the above matrix and C and D play with fixed probability of winning a point (0.529 and 0.471), then A and C have the same chance of winning the game. Note that this solution is also applicable to other situations. For instance, suppose that A after winning a point wins the next point with probability p and, after losing a point, wins the next point with probability q. Then the equivalent game with a fixed probability of winning a point is the one with $x = q/(1 - p + q)$.

Conclusion

Elementary concepts of probability, calculus, and linear algebra are used for analysis of a table tennis game. The modeling is carried out using several different approaches. Comparison of the results under the new and old rules of the game reveals that the outcome is less predictable under the new rules, making the game more exciting. The analysis can be expanded in many different directions and illustrates the steps of modeling.

Rule of Tangent, a Performance Measure

As mentioned earlier, the longest match in professional tennis history took place at Wimbledon in 2010. It lasted eleven hours and five minutes (played over the course of three days), easily eclipsing the previous record of six hours and three minutes set during the 2004 French Open. The Wimbledon match was decided after 980 points, and, most amazingly, the final set (which does not end until one of the players obtains a two-game lead) reached a score of

70–68. The improbability of this score motivates our investigation of win-by-two games.

A game of tennis is won by the first player to obtain at least four points while leading by at least two. A tie score of 3–3 or higher is called deuce. Our main result is the introduction of a trigonometric interpretation of the odds of winning games when serving from deuce. We note that a score of 2–2 is equivalent to deuce in the sense that, in either case, the game continues until one player obtains a two-point lead, so our results apply in this case as well.

In tennis, the same player serves the ball throughout a game, which is generally advantageous for that player, and we suppose that the probability of winning a point is fixed for the duration of the game. One can show that even when one player is significantly better than the other, games will frequently reach deuce or the equivalent state of a 2–2 tie; for example, this will happen over 40 percent of the time when one of the players wins each point with probability 2/3.

A Wimbledon championship match is decided by the best of five sets. However, if the fifth set ends in a tie (6–6), it is extended until one of the players obtains a two-game lead. This is similar to the state of deuce in a game except that the serve alternates from one game to the next.

Numerous aspects of tennis have been studied using both probability and statistical analysis, going back to Bernoulli's *Ars Conjectandi*. Closer to our investigations, the odds of winning a deuce game in tennis were analyzed using geometric series and later by solving a recurrence relation. The same method had been used by Ian Stewart in his 1991 book *Game, Set and Math*. We provide a different perspective on these latter results by showing how the odds of winning a deuce game of tennis can be expressed in simple tennis of trigonometry. We place this result in the more general setting of a gambler's ruin problem and also propose a performance measure to quantify the serving and receiving skill of one player relative to another.

A famous mathematical model known as Markov chains has been applied to the study of sports, most notably baseball, for over fifty years. The probabilities of winning a game, set, and match in tennis were analyzed via Markov chains, and since that time, many authors have used Markov chains to model and reason about various aspects of tennis.

Here, we make the common assumption that points are independent and identically distributed. Klaassen and Magnus (2001) analyzed 86,298 points from Wimbledon matches during the early 1990s and rejected the independent and identically distributed hypothesis, though they noted that the divergence is small enough (especially for strong players) that the hypothesis can be a reasonable approximation of what happens in practice. In any case, we are only analyzing outcomes starting from deuce, where independent and identically distributed points are more expected. We will show that even if the probability of winning a point depends on the state of the game, it is possible in some cases to calculate the odds of winning the game using an equivalent game in which the probability of winning a point is fixed.

A Trigonometric Perspective

Tennis has been analyzed mathematically by a large number of researchers. The rule of tangent introduced in here continues in this direction by translating the analysis of win-by-two games into the language of trigonometry.

We note that most analyses of win-by-two games are carried out in terms of probabilities, but students and the general public are more accustomed to discussing the *odds* of winning. Our rule of tangent expresses the odds of winning a game of tennis when serving from deuce in terms of an angle representing the "climbing difficulty" for one player against another.

This perspective may also lead to new insights or results using the well-established and intuitive properties of trigonometric functions: easily computed derivatives, symmetry, periodicity, etc. Here, we focus on the odds of winning when serving from deuce, but an analysis of more-complicated aspects of win-by-two games or generalizations of such games may lead to cumbersome equations that are more readily understood and simplified when viewed in their equivalent trigonometric form. This point is illustrated in a small way by the obvious fact that if a player wins a game with probability $\sin^2 \theta$, then his opponent wins with probability $\cos^2 \theta$; without the translation to trigonometry, the complementary probability is obtained through simple but tedious algebraic manipulation. Also, the remarks following proposition 1 in the next section illustrate how the first and second derivatives of a trigonometric function can be used to confirm intuition about the odds of winning.

Notation. For $0 < x < 1$, consider the right triangle with legs of length x and $1 - x$. The angle opposite the leg of length x will be denoted by θ. Note that if x is the probability of winning a point on the serve, then the odds of winning a point on the serve are $x/(1 - x) = \tan \theta$.

As mentioned above, we interpret the angle θ as a "climbing difficulty" for the server against the receiver; the steeper the angle, the lower the odds of winning a point on the serve, which, of course, means that the odds of winning the game will be lower.

Tangent Rule

Tennis is the simplest of the games we consider because the same player serves for the duration of a game, and after each rally, a point is awarded to the winner of that rally. We want to calculate the probability $p(x)$ that the server wins a game starting from deuce as a function of the fixed (we assume) probability x that he wins a point. There are two cases to consider:

1. The server wins the first two rallies (and thus the game) with probability x^2.
2. The server and receiver each win one of the first two rallies, which happens with probability $2x(1 - x)$

From these cases, we obtain the relation $p(x) = x^2 + 2x(1-x)p(x)$ which leads to

$$p(x) = \frac{x^2}{x^2 + (1-x)^2} = \sin^2 \theta = \frac{\tan^2 \theta}{1 + \tan^2 \theta}$$

The odds of winning a game when serving from deuce are therefore $\tan^2 \theta$, and we have:

Proposition 1 (rule of tangent for tennis). In a game of tennis, suppose that the server wins each point with a fixed probability x. The odds favoring the server to win a point are, by definition, $\tan \theta$. The odds of winning the game starting from deuce are $\tan^2 \theta$.

Table Tennis

Table tennis (ping-pong) games are played to eleven, and, as in tennis, the winner must obtain a two-point lead. As in tennis, the winner of a rally (whether server or receiver) earns a point. But in table tennis, if a game reaches deuce (10–10), then the serve alternates from one player to the other after each rally. (We noted earlier that a tennis score of 2–2 is equivalent to deuce and that our analysis applied in that case as well, but, here, the analysis applies only to games at deuce because prior to that point, each player gets two consecutive serves.)

We can calculate the probability $p(x, y)$ that player A wins the game when serving from deuce, given the probability x that A wins a point when serving and the probability y that B wins a point when serving. There are three cases to consider:

1. A wins the first two rallies (and thus the game) with probability $x(1 - y)$.
2. Back to deuce: A wins the first rally and B wins the second with probability xy.
3. A loses the first rally and B loses the second with probability $(1 - x)(1 - y)$.

Volleyball

Volleyball differs from tennis and table tennis in that the same team does not necessarily serve each rally and neither does the serve necessarily alternate. Instead, the winner of a rally earns the right to serve the next rally. The winner of a game is the first team to reach a score of twenty-five while leading by at least two. The term *deuce* is not typically used, but we adopt it here to mean any tie score of at least 23–23.

Official volleyball was originally played with side-out scoring. This means that only the serving team can win a point. If the receiving team wins a rally, then it earns only the right to serve the next rally. Rally scoring, introduced in 1999, awards points as in tennis and table tennis—to whichever side wins a rally regardless of whether or not that side served.

Here, we analyze volleyball games under both the old and the new scoring rules. In both cases, we calculate the probability $p(x, y)$ that team A wins the game when serving from deuce, given the probabilities x and y as defined previously.

Equivalent Games

In tennis, the probability of winning a point may change during the match. In some cases, it is possible to analyze the odds of winning the game.

References

Noubary, Reza. 2007. "Analysis of a Table Tennis Game: A Teaching Tool." *IMA Sport*, edited by Percy et al., 147–151.

Franc Klaassen Jan Magnus. 2001, "Are Points in Tennis Independent and Identically Distributed? Evidence from a Dynamic Binary Panel Data Model." *Journal of the American Statistical Association*, vol. 96, 500-509.

Teaching Lessons Using Tennis

During the early years, tennis players used a variety of scoring systems. By the time of the first championship at Wimbledon in 1877, the All England Croquet Club had settled on a scoring system based on the traditions of court tennis. This system remained unchanged until the introduction of tiebreakers in 1970.

As discussed earlier, one quirk of tennis scoring is that strange names are used for points in scoring a game: *love, fifteen, thirty, forty,* and *game.* Although no one knows the origin of this odd system, it has been proposed that *fifteen, thirty, forty-five, sixty* were originally used to represent the four quarters of an hour. Over the years, the score *forty-five* became abbreviated as *forty.* (In informal play, *fifteen* is sometimes abbreviated as *five.*) It would be simpler to score the game *zero, one, two, three, and four.* Still, the weird point names give no advantage to either player.

A more important quirk is that a game must be won by two points. If players each score three points, the score is called *deuce* rather than 40:40. If the server wins the next point, the score becomes *advantage in.* If the server wins again, she wins the game; otherwise, the score returns to deuce. If server loses the next point, the score becomes *advantage out.* If the server loses again, she loses the game; otherwise, the score returns to deuce. This feature of tennis scoring has the virtue of increasing the chance that the stronger player will win, as we shall see.

Consider a game of tennis between two players, A and B. The progression of the game can be used to teach many statistical concepts and critical thinking. Throughout for any event E, we use P(E) to denote the probability that E occurs.

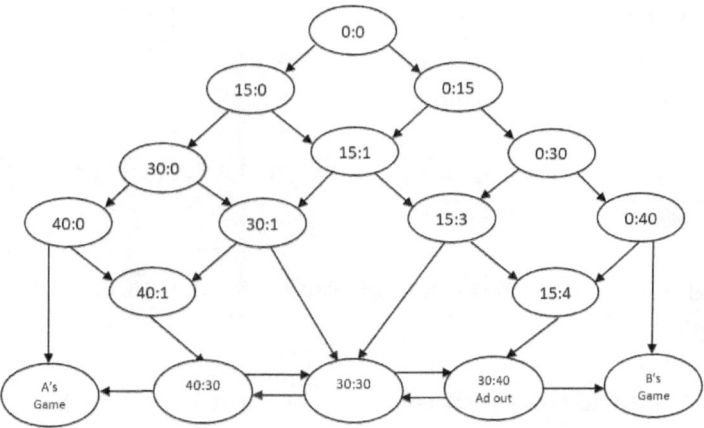

1. Let

$$x = P \text{ (A wins a point)}$$
$$y = 1 - x = P \text{ (B wins a point)}$$

To simplify the modeling and analysis of the game, we may first assume that x is fixed–that is, it does not change during the game. Do you think this is a reasonable assumption?

Objective: To teach critical and logical thinking versus practical significance.

2. Starting from 0:0, what are the possible outcomes after one exchange (one point), two exchanges, and so on?

Objective: To teach concepts such as sample space (all possibilities), events (a collection of outcomes), and their algebra.

3. How do you assign probabilities to the outcomes of the sample space after one exchange? Note that possible outcomes are 0:15 and 15:0.

Objective: To teach concepts such as quantification of uncertainty, probability (classical, objective, and subjective), and odds.

4. Do you think winning a point will affect the probability of winning the next point?

Objective: To teach concepts such as conditional probability and independence.

5. Suppose that the answer to question 4 is no. Let $x = 0.60$. How do you find the probabilities such as P (15:15) or P (30:30) or P (40:15), etc.? How do you find these probabilities if the answer to question 4 is yes?

Objective: To teach combinations, multiplication rule, Bernoulli, binomial, and Poisson distributions.

6. Find (*a*) P (A wins the game without deuce),
       (*b*) P (A wins the game after reaching deuce),
       (*c*) P (A wins the game)

Objective: To teach addition rule, infinite series, and geometric progression.

7. Find (*a*) P (A wins the set without tie-break),
       (*b*) P (A wins the set after a tie-break),
       (*c*) P (A wins the set)
   Objective: To teach modeling and problem solving.

8. Find P (A wins the match)

Objective: To teach pattern identification and model building.

9. Find general formulas for probabilities in questions 6, 7, and 8.

   Objective: To teach functions, graphs, and function of functions (composite functions).

10. Let $e = x - y$ represent the edge in one point. For example, $x = 0.51$ means player A has a 0.02 edge over player B in one point. Find edges in one game, one set, and the match.

    Hint: call the edge in one point Δx and edge in a game Δy.

    Objective: To teach the derivative, chain rule, and differential equations.

11. Let $r = x/y$. Since $0 \le x, y \le 1$, it follows that $0 \le r < \infty$ and $r = 1$ for $x = y = 0.5$. Express the probabilities in question 9 in terms of r.

    Objective: To teach transformations and homogeneous polynomials.

12. Let 0, 1, 2, 3, and 4 represent the scores 0, 15, 30, 40, and the game, respectively. Let $g(i,j) = $ P (A wins the game starting from the score $(i:j)$). Show that

$$g(i,j) = xg(i+1,j) + yg(i,j+1).$$

    Also show that $g(3,3) = x^2/(x^2 + y^2)$.

    Objective: To teach recursions and difference equations.

13. Think about a game that has reached the state deuce (or 30:30). Recall that there is no limit to how long the game

could go on. From this point, the game could reach one of the five possible states. Let 1, 2, 3, 4, and 5 denote the states: A's game, B's game, deuce, advantage A, and advantage B, respectively. Recall that the game moves from state to state until one player wins. The probabilities of moving from one state to another can be summarized as

$$\begin{array}{c} \\ 1 \\ 2 \\ 3 \\ 4 \\ 5 \end{array}\begin{array}{ccccc} 1 & 2 & 3 & 4 & 5 \\ \begin{bmatrix} 1 & 0 & 0 & 0 & 0 \\ 0 & 1 & 0 & 0 & 0 \\ 0 & 0 & 0 & x & y \\ x & 0 & y & 0 & 0 \\ 0 & y & x & 0 & 0 \end{bmatrix} \end{array}$$

Objective: To teach matrices, Markov chains, and states of a Markov chain.

14. Suppose that the game is now in state deuce (state 3). This can be expressed as the state matrix:

$$\begin{array}{ccccc} 1 & 2 & 3 & 4 & 5 \end{array}$$
$$\begin{bmatrix} 0 & 0 & 1 & 0 & 0 \end{bmatrix}$$

15. Show that after one and two exchanges, the state matrices are respectively

$$\begin{array}{ccccc} 1 & 2 & 3 & 4 & 5 \end{array} \qquad \begin{array}{ccccc} 1 & 2 & 3 & 4 & 5 \end{array}$$
$$\begin{bmatrix} 0 & 0 & 0 & x & y \end{bmatrix} \qquad \begin{bmatrix} x^2 & y^2 & 2xy & 0 & 0 \end{bmatrix}$$

Objective: To teach matrix algebra.

16. Starting from deuce,

a. How many exchanges (points) are expected to be played before the game ends?
b. How many times is each state expected to be visited/revisited before the game ends?

Objective: To teach stationary solution, inverse of a matrix, and the fundamental matrix.

17. Suppose now that x_1, x_2 represent respectively the probabilities in part a and x_3, x_4 represent respectively the probabilities in part b.

   a. P(A wins a point when serving) and P(A wins a point when receiving)
   b. P(A wins a point after winning a point) and P(A wins a point after losing a point)

Find the probabilities of winning a game, a set, and the match for player A.

Objective: To teach basic concepts of modeling.

18. Consider a tournament like the Davis Cup. Suppose that countries A and B each have three players represented as A_1, A_2, A_3 and B_1, B_2, B_3, respectively. Suppose that the following matrix represents their chances of winning or losing against each other.

$$
\begin{array}{c@{\qquad}ccc}
& B_1 & B_2 & B_3 \\
\begin{array}{c} A_1 \\ A_2 \\ A_3 \end{array} &
\left[\begin{array}{ccc}
40\% & 52\% & 50\% \\
40\% & 41\% & 30\% \\
55\% & 45\% & 60\%
\end{array}\right]
\end{array}
$$

For example, using this matrix, we have $P(A_1 \text{ beats } B_1) = 40\%$. Recall that in the Davis Cup, each team decides which player plays the first, second, etc., game without knowing about the selection of the other team. How do you think teams should make their selection?

Objective: To teach game theory.

19. Recall that in tennis, the server gets a second chance to serve after missing the first one. Ordinarily, players go for a speedy (strong) but risky first serve and a slow but a more conservative second serve. Analyze all the possible serving strategies and their consequences.

Objective: To teach basic concepts of decision analysis and its role in the game theory.

20. How do you summarize statistics related to a tennis player, a team, and a tournament?

Objective: To teach descriptive statistics.

21. Suppose that you have data for the speed of player A's first serve. How do you calculate the probability that in the next match, the average speed of A's first serves would exceed a certain value?

Objective: To teach sampling distribution and central limit theorem.

22. How do you compare two tennis players? How do you rank tennis players?

Objective: To teach performance measures, measures of relative standing, z-score, etc.

23. A claim is made about the performance of a tennis player. Using the player's statistics, how do you validate the claim?

    Objective: To teach hypothesis testing, type I and type II errors and P-value.

24. How can you use the past statistics of a player to predict his/her future performance?

    Objective: To teach estimation (prediction), confidence intervals, regression, time series, and forecasting.

25. Suppose that you have statistics on the speed of player A's first serves. How do you predict the next record speed and perhaps the maximum possible speed of A's serves?

    Objective: To teach theory of records, asymptotic theory of order statistics, extreme value theory, and threshold theory.

26. How do you organize a tennis tournament?

    Objective: To teach planning and scheduling.

27. Recall that the winner of men's tennis match must win three out of five sets. Each set has six games. Do you think the present scoring system is fair? For example, player A could win two sets six games to none (6–0) and lose three tiebreak sets (6–7). In this case, A could win thirty games and lose only twenty-one games and yet lose the match. Do you have any suggestion to make the match more balanced?

    Objective: To teach methods for adaptive modeling.

Activities

We conclude with some activities that use tennis and certain mathematical and statistical concepts. For example, since $0 < x < 1$, we may replace it with $\sin 2\alpha$. Then $y = 1 - \sin 2\alpha = \cos 2\alpha$. Using this, we can develop relate above calculations to well-known concepts of trigonometry.

Binomial Distribution, Matrices, Markov Chain, and Derivatives

Consider a match between two players, A and B. Suppose that player A has a 10 percent edge over player B in one point–that is, the probability that player A will win a point is 55 percent, and the probability that player B will win a point is 45 percent.

1. Show that the probability calculations before reaching deuce can be carried out using a binomial distribution.
2. Find the probability that player A wins a game, a set, and the match given the edge A has in one point. Also calculate the edge in a game, a set, and a match given the edge in one point.

Suppose that the information regarding the players A and B is summarized in a rectangular array given below.

$$
\begin{array}{c c c c c c}
 & 1 & 2 & 3 & 4 & 5 \\
1 & \begin{bmatrix} 1 \\ 0.55 \\ 0 \\ 0 \\ 0 \end{bmatrix} & \begin{matrix} 0 \\ 0 \\ 0.55 \\ 0 \\ 0 \end{matrix} & \begin{matrix} 0 \\ 0.45 \\ 0 \\ 0.55 \\ 0 \end{matrix} & \begin{matrix} 0 \\ 0 \\ 0.45 \\ 0 \\ 0 \end{matrix} & \begin{matrix} 0 \\ 0 \\ 0 \\ 0.45 \\ 1 \end{matrix}
\end{array}
$$

This is called a transition matrix. It includes the probabilities of moving from one state to another after one point. Here, state 1 represents A won the game, state 2 represents the advantage A, state 3 represents the deuce, state 4 represents the advantage B, and state 5 is B won the game.

3. Apply matrix algebra and interpret the results in the context of a tennis game.
4. If we look at the tennis game as a Markov chain, what are the states? Which states are nonrecurrent? Which states are recurrent? Which states are absorbing?
5. Let x denote the probability that player A wins a point, and $y = 1 - x$ denote the probability that player B wins that point. It can be shown that

$$P \text{ (A wins the game)} = x^4[1 + 4y + 10y^2 + \frac{20xy^3}{x^2+y^2}]$$

Replace y by $1 - x$ in this formula and find its derivative with respect to x. Calculate the value of the derivative at the point $x = 0.50$. For players close in ability (small edge in one point, e.g., 1 percent), the resulting value provides the edge in one game. Compare the value obtained using derivative with the actual value of the edge.

6. Consider the formula in problem 5. Replace y by $1 - x$. The resulting function has several properties. For example, it is symmetric with respect to $x = 0.5$. Study the other properties of this function.
7. Consider the formula in problem 5. Replace y by $1 - x$. Suppose that P (A wins the game) $= 0.60$. Use numerical method to find x.
8. Find the probability of winning a set as a function of x and show that this is an example of a function of a function. Use this to find the edge in a set both directly and by using the derivative (chain rule) as in problem 5.

Calculations Based on Normal Distribution

1. The average speed of a well-known tennis player's first serve is 117 mph with a standard deviation of 5 mph. What is the probability that this player's next first serve

   (a) will be slower than 115 mph?
   (b) will be faster than 120 mph?
   (c) would have a speed between 116 mph and 122 mph?

2. Suppose that tennis balls are produced to have COR = 55.5 percent (target value). To see if the process is on target, once a day, fifty balls are tested. If the average COR falls outside the interval 53.5 and 57.5 percent, the process is judged out of control. What is the probability that the process will be judged out of control incorrectly? Assume that the standard deviation is 1.5 percent.

Confidence Intervals and Testing Hypotheses

1. Look up statistics for the number of aces made in thirty-six matches by a tennis player of your choice. Construct a 95 percent confidence interval for the average number of aces for this player. Hint: Use the central limit theorem.
2. A sample of thirty-six serves of a top player on hard court has mean of 107 mph and standard deviation of 6.5 mph. His coach claims that the average speed of his serves is 110 mph. Check to see if data supports this claim. Use a 0.05 level of significance.
3. Another sample of thirty-six serves from the same player (problem 2) on a grass court has mean of 105 mph and standard deviation of 10 mph. Can we conclude that population mean speed of his serves on the hard court is

greater (at the 0.05 level) than the population mean speed of his serves on the grass court?

4. Construct a confidence interval for the difference between the population means in problems 2 and 3.

Regression and Time Series

1. Use statistics for a player of your choice who did participate in the latest Wimbledon tournament. Use regression to predict the total points won using, for example, the number of opponents' unforced errors as predictor. Try other factors to see if you can find the best predictors.

2. Recall that if we cannot apply regression, we can still use smoothing techniques for prediction when the data form a time series. Time series refers to data with a time index. Use smoothing to predict number of matches a player of your choice may win next year using the number of matches the player has won in previous years.

Research topics

The analysis of a tennis game can be expanded in many different directions. Examples include Canadian doubles or cutthroat, Australian doubles, regular doubles, and even a five-way game when server sits out for a game. Examples of research topics include determination of the size of a handicap to make a game a fair game and analysis of the methods used for ranking tennis players.

References

Collings, B. J. 2007. "Tennis (and Volleyball) without Geometric Series." *College Mathematics Journal*, 38, 55–57.

Croucher, J. S. 1986. "The Conditional Probability of Winning Games of Tennis." *Research Quarterly for Exercise and Sport*, 47, 23–26.

Fischer, G. 1980. "Exercise in Probability and Statistics, or the Probability of Winning at Tennis." *American Journal of Physics*, 48, 14–19.

Gale, D. 1971. "Optimal Strategy for Serving in Tennis." *Mathematics Magazine*, 44, 197–199.

Gallian, J. A. (ed.) 2010. *Mathematics and Sports*. Mathematical Association of America.

Gillman, L. 1985. "Missing More Serves May Win More Points." *Mathematics Magazine*, 58, 222–224.

Gould, R. J. 2010. *Mathematics in Games, Sports, and Gambling: The Games People Play*. CRC Press.

Klaassen, F. J., and J. R. Magnus. 2001. "Are Points in Tennis Independent and Identically Distributed? Evidence from a Dynamic Binary Panel Data Model." *Journal of the American Statistical Association*, 96, 500–509.

Newton, P. K., and K. Aslam. 2009. "Monte Carlo Tennis: A Stochastic Markov Chain Model." *Journal of Quantitative Analysis in Sports*, 5.

Newton, P. K., and J. B. Keller. 2005. "Probability of Winning at Tennis I. Theory and Data." *Studies in Applied Mathematics*, 114, 241–269.

Noubary, R. 2007. "Probabilistic Analysis of a Table Tennis Game." *Journal of Quantitative Analysis in Sports*, 3.

Noubary, R. 2010. "Teaching Mathematics and Statistics using Tennis." Number 43 in *Dolciani Mathematical Expositions*. Mathematical Association of America, chapter 19, 227–239.

O'Malley, A. J. 2008. "Probability Formulas and Statistical Analysis in Tennis." *Journal of Quantitative Analysis in Sports*, 4.

Sandefur, J. 2005. "A Geometric Series from Tennis." *College Mathematics Journal*, 36, 224–226.

Snell, J. L. 1959. "Finite Markov Chains and Their Applications." *American Mathematical Monthly,* 66, 99–104.

Winston, W. L. 2009. *Mathletics.* Princeton University Press.

Wong, R., and M. Zigarovich. 2007. "Tennis with Markov." *College Mathematics Journal,* 38, 53–55.

CHAPTER 4

Rare Performances and Records

I N SPORT, EXTRAORDINARY performances are the ones that do significantly better than expectation and pass a threshold for ordinary. For example, in one-hundred-meter races, only male athletes who run the distance in less than ten seconds may be considered exceptional. In basketball, a player with a point per minute or assist per minute greater than 2/3 may be referred to as exceptional. In tennis, players having more than forty-five aces in one match may be considered exceptional. If we examine the distribution of times for one-hundred-meter races for the runners or points per minute (PPM) for basketball players, then we find that the exceptional performances constitute the tail of the corresponding distribution (for one-hundred-meter times, the lower tail; and for PPM, the upper tail). Thus, to study the records or to predict them, rather than the entire distribution covering the entire range of possible values, we may just study the upper (lower) tail of the corresponding distribution. We can do this either by seeking a model for values above (below) a specified threshold or by considering a past performance (e.g., the third best performance) in the history of the sport of interest and study the number of times it was surpassed or exceeded.

Exceedances are used to answer questions such as, how many (basketball players) would it take to produce a player that will perform like or better than, for example, Michael Jordan?

In many situations in sports, instead of records, we are interested in the events associated with the exceedances of certain performance measures of the variable under study. For example, in sports such as long jump, one needs to jump a certain length to qualify for a competition. This leads to deal with the frequencies (number of athletes) instead of the values on the random variable itself. For instance, we may like to know how many baseball players will have more than fifty homeruns next season. Or how many of the players will surpass the performance of the third-best player in the history of a given sport? Many authors have discussed exceedances and their applications.

In sum, the celebrated central limit theorem has given statistics its focus on averages—for we do what we know how to do. The theories of extremes and record values are less simple, less unified, and more recent—but not less important. The question then arises as to whether there is a similar limiting theorem for maximum, minimum, and record values as for the mean.

Analysis of Extremes/Major Methods

Out-of-ordinary/extreme/outstanding performances in sports are often analyzed using one of the three available methods itemized below together with the *theory of exceedances* that deals exclusively with the number of times a chosen threshold is exceeded.

1. The *extreme value theory*, which usually deals with the maxima or minima. The method uses the absolute largest or absolute smallest values in a specific period.
2. The *threshold theory*, which deals with values above or below a specified threshold.
3. The *theory of records*, which deals with values larger or smaller than all the previous values.

Extreme Value Theory, Threshold Theory, Theory of Records

Extreme value theory

Extreme value theory deals with the annual (or any other period) maxima or minima. Specifically, the theory is based on dividing the sample into subsamples and fitting a distribution for maxima or minima of the subsamples. For example, data may consist of largest earthquakes in California for each of the last one hundred years. Or data may consist of best time of one-hundred-meter run for each of the last one hundred years.

The forms of the limiting distributions are specified by the extreme value theorem. This theorem states that there are three possible types of limiting distributions:

1. The Gumbel distribution (type I)
2. The Frèchet distribution (type II)
3. The Weibull distribution (type III)

These three forms can be combined to yield the so-called generalized extreme value distribution (GEV). Basic results are the following:

- Only distributions unbounded can have a Frèchet distribution as a limit.

- Only distributions with finite endpoint can have a Weibull as a limit.
- The Gumbel distribution can be the limit of bounded or unbounded distributions.

Note that since extreme value distributions are fitted to, for example, annual maxima or minima, some relevant data related to the years with several large observed values could be discarded, and some less informative data related to the years with no real large values could be retained.

Threshold theory

The threshold theory allows one to make inference about the values above or below a threshold–that is, the upper or the lower tails of a distribution. It considers the excesses, the differences between the observations over the threshold, and the threshold itself. Like extreme value distributions, there are three models for tails:

1. long-tail Pareto
2. medium-tail exponential
3. short-tail distribution with an endpoint

These three forms can be combined to yield the so-called generalized Pareto distribution (GPD). Again, most classical distributions fall in domain of attraction of one these tail models. It has been shown that the natural parametric family of distributions to consider for excesses (tail) is the generalized Pareto distribution (GPD). Note that only distributions unbounded can have a Pareto or exponential distribution as a limit.

The theory is very useful when modeling large values based on observed large values, which is of main concern.

Theory of records

Theory of records deals with values that are strictly greater than or less than all previous values. Usually, the first value is counted a record. Then a value is a record (upper record or record high) if it is bigger than all previous values.

The study of record values, their frequencies, times of their occurrences, their distances from each other, etc., constitutes the theory of records. Formally, the theory deals with four main random variables:

1. the number of records in a sequence of n observations
2. the record times (times records occur)
3. the waiting time between the records
4. the record values

It is interesting to note that the first three can be investigated using nonparametric or distribution-free methods (applicable regardless of underlying distribution), whereas the last one requires parametric methods. The theory is developed for independent and identically distributed observation, which makes it problematic to apply to sport data.

Because of the lack of independence and a few other factors, some of which are listed below, sport records break more often than what the theory of records predicts:

1. improvement in training, coaching, equipment, diet, etc.
2. increase in participation or attempts
3. participation of same athletes (e.g., runners)

To account for this, some adjustments are therefore necessary. We can treat the problem as if either participation has increased with time or more competitions have taken place so that the chance of setting a new record was increased. To clarify, suppose that the probability of breaking the record is p for n participants.

Suppose further the participation is increased to $3n$. Let us divide participants into three groups of n participants. Then the probability that at least one group break the record is

$$1 - (1 - p)^3.$$

For example, this probability is 0.143 for $p = 0.05$ and 0.271 for $p = 0.10$.

Theory of Exceedances

Theory of exceedances deals with number of times a threshold is exceeded. As a result, it is a counting process. Some examples in sports include the following:

1. In track and field, athletes need to exceed specified qualifying levels (e.g., time, distance, or height) to be selected for certain competitions. We are concerned with the number that succeed. For instance, in the long jump, a distance of eighteen feet, nine inches will qualify for districts.
2. In sports such as basketball, we may be interested in the number of players who average more than m points per game.

Let Y be a random variable representing, for example, a performance measure in a given sport and x a real number representing the qualifying level. The event $Y = y$ is an exceedance of the level x if $y > x$. Now, assuming independent and identically distributed trials, what is the probability of r exceedances in the next n trials?

This is clearly a Bernoulli experiment with two possible outcomes: "exceedance" or "not exceedance." Since it is repeated n times, the number of exceedances has a binomial distribution with parameters; n, $p(x)$, where $p(x)$ is the probability of exceedance of the level x of the variable under study. Note that, here, $p(x) = P(X > x) = 1 - F(x)$, where $F(x)$ is the distribution function of X. Using this, the probability of r exceedances of level x in the next n trials can be calculated from

$$\binom{n}{r}[1 - F(x)]^{r} F^{n-r}(x) \,, 0 \le r \le n$$

Now, suppose that, rather than a fixed level, we make the level x dependent on, say, n, x_n. This means that we change the level with increase in frequency of events such as earthquakes. If we choose x_n in such a way that the following condition is satisfied:

$$\lim_{n \to \infty} n[1 - F(x_n)] = \tau, \quad 0 \le \tau \le \infty$$

then the probabilities of r exceedances of level x_n can be approximated by those of a Poisson distribution (process) with parameter (rate) τ. This is because exceedances become rare events and the binomial distribution tends to Poisson.

Another interesting problem related to exceedances is the following: Assuming independent and identically distributed trials, determine the probability distribution of number of exceedances in the next N trials of the mth largest observation in the past n trials. This is useful when choosing a design load smaller than magnitude of some that have already occurred.

Suppose that p_m is the probability of exceedance of the mth largest observation in the past n trials, then it can be shown that pm has Beta distribution with density function

$$f(p_m) = \frac{p_m^{m-1}(1 - p_m)^{n-m}}{B(m, n-m+1)}; \quad 0 \le p_m \le 1$$

Using this, and the fact that the probability of r exceedances in the next N trial is binomial with parameters N and p_m, it can be shown that the mean and the variance of the number of exceedances of the mth largest observation in future N trials are respectively

$$\frac{Nm}{n+1}, \quad \frac{Nm(n-m+1)(N+n+1)}{(n+1)^2(n+2)}$$

Example. The yearly maximum number of points scored in high school basketball games during the last forty years was seventy. Then the mean and variance of the number of exceedances (details is discussed later) of seventy of the yearly maximum number of the points during the next thirty years are respectively

$$30/41 = 0.732$$
$$(30)(40)(71)/(41)^2(42) = 1.207$$

If the second-largest score was sixty-seven, then the mean and variance of the number of exceedances of this value in the next twenty years are respectively

$$(20)(2)/41 = 40/41 = 0.976$$
$$(20)(2)(39)(61)/(41)2(42) = 1.348.$$

Example. Consider the yearly maximum score during the last sixty seasons. As a goal in our school, we want to choose a level to have a mean value of four exceedances in the next twenty years.

Using the formula for the mean, we get $20m/61 \approx 4 \Rightarrow m \approx 12$. This means that the value to be chosen is the twelfth-largest score (order statistic) in the data.

Now, suppose that m=1. It follows from (1) that

$$f(p_1) = n(1-p_1)^{n-1}; \quad 0 \le p_1 \le 1$$

For this distribution, the mean and variance are respectively

$$\frac{1}{n+1}, \quad \frac{n}{(n+1)^2(n+2)}$$

Noting that for a relatively large n, the variance is small (e.g., for fifty years of data, it is equal to 0.0003696), we may replace p1 with its expected value so that the probability of exceedances in the next trial is $\frac{1}{n+1}$. Using this, the probability of no exceedances in the next N trials is

$$(1-\frac{1}{n+1})^N$$

Example. Recall the basketball scores mentioned above. We have $n - 40$ and $N - 30$. Thus, the required probability is $(1 - (1/41))^{30} = 0.4767$, and the probability of at least one exceedances is 0.5233.

Note that a good approximation for $(1-\frac{1}{n+1})^N$ can be found if we let $N = h(n + 1)$. This gives

$$[(1-\frac{1}{n+1})^{\frac{1}{n+1}}]^h \rightarrow \exp(-h) = \exp(-\frac{N}{n+1})$$

In the above example, h = 30/41 and exp(-30/41) = 0.481.

The above approximation was based on replacing the expected value for the probability of no exceedance. Since the distribution of p_1 is not symmetric, one may prefer to replace the mean (expected value) with the median—that is, consider $1-2^{-\frac{1}{n}}$ in place of $\frac{1}{n+1}$. This gives $2^{-\frac{N}{n}}$ in place of $(1-\frac{1}{n+1})^N = \exp(-\frac{N}{n+1})$. Continuing with the basketball example, $2^{-\frac{30}{40}} = 0.595$.

Finally, suppose that K is the number of trials up to the first exceedance. The possible values for K are 0, 1, 2, ..., and we have

$$P(K = k) = \frac{1}{n+1}(1 - \frac{1}{n+1})^{k}, \ k = 0, \ 1, \ 2, \ \ldots$$

This is a geometric distribution with expected value and variance equal to respectively

$$n+1, \ n(n+1).$$

This means that to have an exceedance, we need an average of n+1 trial.

Exceedances and English Premier League

Astrophysicists at the University of Warwick studying the extreme variability in X-rays emitted from matter falling into black holes have discovered that their research methods also show that the world's top division football (soccer) matches have an unusually large proportion of high-scoring games—so much so that international football (soccer) actually shows a pattern of "extreme events" similar to that seen in the large bursts of X-rays from the accretion discs of black holes. However, analysis of just English premier football (soccer) league and cup games showed that English top division football (soccer) is in fact thirty times less likely to have high-scoring games than the rest of the world taken as a whole and could thus be seen by some people as thirty times more boring.

"External events"—that is, large events that are more likely than would be expected from a random process—can be a signature of complexity in nature. In the case of matter moving in accretion disks around black holes, it tells us about the turbulent flow in the accreting matter.

While seeking to compare this distribution of events with other patterns in the world around us, two University of Warwick postgraduate physics students (John Greenhough and Paul Birch), with their supervisor Sandra Chapman and colleague George

Rowlands, looked at the number of goals scored by the home and away teams in over 135,000 games in 169 countries since 1999 and found that the results followed the pattern of "external statistics" (https://www.eurekalert.org/pub_releases/2001-11/uow-bhr112601.php).

However, they also compared their data with an analysis they made of the scores of thirteen thousand English top division games and five thousand FA Cup matches between the 1970/71 and 2000/01 seasons. They found that these scores contained far less high-scoring games than the world as a whole, and rather than fitting an external statistics pattern, the English games more closely fitted either Poisson or negative binomial distributions.

In summary, their analysis revealed that a total score over one hundred goals in any one game occurs approximately only once in every ten thousand English top division matches (once every thirty years), but in top division matches worldwide, such a score is seen once in three hundred games (about once every day).

References

Getachew, Abraham. "Traffic Load Effects on Bridges Statistical Analysis of Collected and Monte Carlo Simulated Vehicle Data." http://citeseerx.ist.psu.edu/viewdoc/download?doi=10.1.1.13.5299&rep=rep1&type=pdf.

Return Periods

Assume an event (breaking a record, winning the championship, etc.) such that its probability of occurrence in a unit period of time (normally one year) is p. Assume also that occurrences of such an event in different periods are independent. Then, as time passes, we have a sequence of equally likely Bernoulli/Binary experiments—only two possible outcomes: (a) occurrence or (b) not occurrence. Thus, the time (measured in unit periods) to the first occurrence is a geometric random variable with parameter

p with expected value/mean value of $1/p$. This motivates the following definition.

Definition: Let A be an event, and T the random time between consecutive occurrences of A events. The mean value, τ of the random variable T is called the return period of the event A.

"Traffic Load Effects on Bridges Statistical Analysis of Collected and Monte Carlo Simulated Vehicle Data" by Abraham Getachew says to note that if $F(x)$ is the distribution function of the yearly maximum of a random variable, the return period of that random variable to exceed the value x is $1/[1 - F(x)]$ years. Similarly, if $F(x)$ is the distribution function of the yearly minimum of a random variable, the return period of the variable to go below the value x is $1/F(x)$ years.

Getachew also says to note that if a given engineering work fails when, and only when, the event A occurs, its mean lifetime coincides with the return period of A. The importance of return periods in engineering is because many design criteria are defined in terms of return periods.

The probability of occurrence of the event A before the return period is (see the geometric distribution)

$$F(\tau) = 1 - (1 - p)^{\tau} = 1 - (1 - p)^{1/p}$$

which for $\tau \to \infty (p \to 0)$ tends to the value 0.63212.

Example. One very interesting example for this is the return period for a player like Michael Jordan. In section 3.6, it was shown the distribution of points per minute for guards who played in 92–93 season was normal with mean 0.4236 and standard deviation 0.1159. Using this, we found that for Jordan, we have

$$P(PPM > 0.8291) \approx 1/5000$$

This means that the return period for a player who would perform like or better than Jordan is five thousand players. So if one hundred guards are added to the NBA each year, on average, it takes about fifty years to produce such a player.

Example . Suppose that the distribution function of the yearly maximum number of points scored in basketball in a given region is given by

$$F(x) = \exp[-\exp\left(-\frac{x - 38.5}{7.8}\right)]$$

Then the return periods of yearly maximum score of sixty and seventy points are respectively

$$\tau_{60} = \frac{1}{1 - F(60)} = 16.25 \text{ years, and}$$

$$\tau_{70} = \frac{1}{1 - F(70)} = 57.24 \text{ years,}$$

This means that the maximum score of sixty and seventy points occur, on average, once every 16.25 ($\approx 16 - 17$) and 57.24 ($\approx 57 - 58$) years, respectively.

Example. Suppose that in a certain tournament, an exceptional performance is defined as that value with a return period of fifty years, and the yearly maximum number of the points scored is known, from past experience, to have a limiting Gumbel distribution with distribution function

$$F(x) = \exp[-\exp(-\frac{x-15}{4})]$$

Then the number of the points h must satisfy the equation

$$\frac{1}{1 - F(h)} = 50$$

from which we get $h = 30.61 \approx 31$ points.

Threshold Theory

In this approach, the probabilities of future performances are calculated by developing models for a tail of the distribution for performance measures. Since performance measures above or below a threshold carry more information about the future exceptional performances, methods based on tail modeling are appropriate. Methods based on this approach assume that the tail of the distribution for the performance measure of interest belongs to a parametric family and carries out the inference using excesses–that is, the performance measures greater or smaller than some predetermined value y_0. It has been shown that the natural parametric family of distributions to consider for excesses is the generalized Pareto distribution (GPD),

$$P(Y \leq y) = 1 - (1 - \tfrac{ky}{\sigma})^{1/k}, \, y \leq {}^{\sigma}/k$$

where Y represents a performance measure and $\sigma > 0$ and $-\infty < k < \infty$ are unknown parameters. The range of Y is $0 < y < \infty$, for $k \leq 0$, and $0 < y < \sigma/k$ for $k > 0$. For example, for the men's one-hundred-meter dash, a threshold such as ten seconds may be considered.

The GPD includes three specific forms:

1. long-tail Pareto distribution
2. medium-tail exponential distribution
3. short-tail distribution with an endpoint

Most classical distributions have tails that behave like one of these forms. Unfortunately, like most asymptotic results, applying this approach is not free of problems. The obvious problems are the choice of a parametric family, determination of the threshold, and having to deal with intractable equations that need to be solved to obtain estimates of the parameters. Hill (1975) has proposed an approach that is easy to use and is applicable to a wide class of distribution functions possessing medium or long tails. They propose assuming a tail model of the form $F(y) = cy^{-a}$, $y > y_0$. It is a suitable model for men's one-hundred-meter data considering that the recent records set by Bolt indicate a long tail.

From a random sample; $Y_1, Y_2, \ldots, Y_n$, the estimates of the parameters are obtained based on the upper $m = m(n)$ order statistics (best times), where m is a sequence of integers chosen such that $m \to \infty$ and $m/n \to 0$. Here, c is estimated using the empirical $1 - m/n$ quantile, $Y_{(m+1)}$ and $1/a$ is estimated by

$$\hat{a}^*(n/m) = m^{-1}\sum_{i=1}^{m} \ln Y_{(i)} - \ln Y_{(m-1)}$$

Statistical theory regarding these estimators is well established. The Pareto-tail upper tail estimate $\bar{F}(y)$ is then

$$g(x) = \sum_{j=0}^{9}\binom{10+j}{10}x^{11}(1-x)^j +\binom{20}{10} x^{12}(1-x)^{10}\big/(x^2 +(1-x)^2)$$

The lower tail estimate can be obtained similarly.

References

Hill, B. M. 1975. "A Simple General Approach to Inference about the Tail of a Distribution." *Annals of Statistics,* 3, 1163-1174.

Theory of Records

Records occur everywhere from sports to the stock market. *The Guinness Book of World Records* is popular reading around the world. The mathematics behind the theory of records is both interesting and elegant.

Consider men's long jump data for 1962–1991. The set contains distances corresponding to thirty years and has five upper records. Records (upper) refer to the distances (values) that are *strictly larger* than the previous distances. Theory of records attempts to find answers to the following questions:

1. How many records do we expect to observe in n attempts?
2. Time interval (waiting time) between records.

Year	Distance (m)	Year	Distance (m)	Year	Distance (m)
1962	8.31 *	1973	8.34	1984	8.71
1963	8.30	1974	8.30	1985	8.62
1964	8.34 *	1975	8.45	1986	8.61
1965	8.35 *	1976	8.35	1987	8.86
1966	8.33	1977	8.33	1988	8.76
1967	8.35	1978	8.32	1989	8.70
1968	8.90 *	1979	8.52	1990	8.66
1969	8.34	1980	8.54	1991	8.95 *
1970	8.35	1981	8.62		
1971	8.34	1982	8.76		
1972	8.34	1983	8.79		

To answer these questions, we begin by letting

$$Y_i = 1, \text{ if } X_i \text{ is a record;}$$
$$0, \text{ if } X_i \text{ is not a record.}$$

From the table of winning long jump distances, we have

X_i: 8.31, 8.30, 8.34, 8.35, 8.33, 8.35, 8.90, 8.34, 8.35, 8.34, 8.34, 8.34, 8.34
Y_i: 1,   0,   1,   1,   0,   0,   1,   0,   0,   0,   0,   0,   0
X_i: 8.30, 8.45, 8.35, 8.35, 8.32, 8.52, 8.54, 8.62, 8.76, 8.79, 8.71, 8.62, 8.61,
Y_i: 0,   0,   0,   0,   0,   0,   0,   0,   0,   0,   0,   0,   0
X_i: 8.86, 8.76, 8.70, 8.66, 8.95
Y_i: 0,   0,   0,   0,   1

Noting that all X_i's are equally likely to be the maximum;

$$P(Y_i = 1) = P(X_i \text{ is a record}) = 1/i$$

This is because the first i values can be arranged in $i!$ ways. If we place the largest among these i values in the ith place, the remaining $i - 1$ values can be arranged in $(i - 1)!$ ways.

Thus, the probability that X_i is a record is $(i - 1)!/i! = 1/i$.

$$E(Y_i) = 0(1 - 1/i) + 1(1/i) = 1/i$$

$R_{n:}$ the number of records in a sequence of length n, then

$$R_n = Y_1 + Y_2 + ...+ Y_i + ... + Y_n = \sum_{}^{n} Y_i$$

since the Y values are either 1 or 0. In fact, here we have

Term	Expected number of Records
1	1
2	1/2
3	1/3
4	1/4
$\vdots$	$\vdots$
n	1/n

This is because the probability that the first number is a record is always 1.

Then the second of two numbers in random sequence has 50 percent chance (or 1/2) that it will be a new record high surpassing the initial record.

Next, the probability that the third number will be a new record is 1/3, since the value must be either the smallest, middle, or the largest.

Therefore, the average number of record highs in R_n, in a sequence of n independent observations is the sum of all of the probabilities—that is,

$$R_n = Y_1 + Y_2 + ... + Y_i + ... + Y_n = \sum_{i}^{n} Y_i$$

$$E(R_n) = 1 + 1/2 + 1/3 + ... + 1/n = \sum_{i=1}^{n} 1/i$$

It can be shown that as $n \to \infty$, $E(R_n)$ approaches $\ln(n) + \gamma$, where $\gamma = 0.5772$ (Euler's constant).

Expected # of Records (R_n)	# of Attempts Years (n)
1.83	3
2.45	6
2.83	9
3.10	12
3.32	15
3.50	18
3.65	21
3.78	24
3.89	27
3.99	30
4.08	33
4.17	36
4.25	39
4.35	43
4.50	50
4.67	60
4.80	70
4.96	80
4.97	81
4.99	82
5.00	83
5.01	84
5.03	85

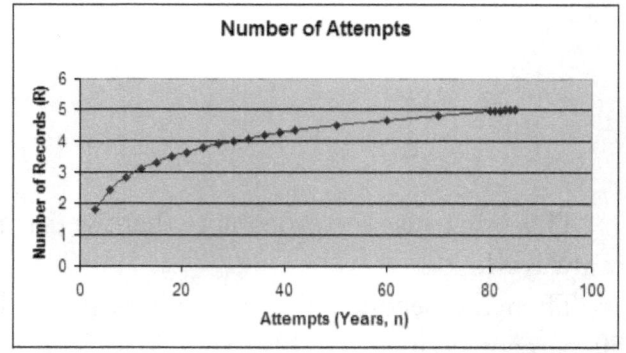

Record Times

Most of the results of theory of records are nonparametric or distribution-free. This makes such results interesting and useful. For example, the probability that the second record occurs in the *jth* place is

$$P(N_2 = j) = 1(j(j-1)), \quad j \geq 2$$

Waiting Times between Records

Although the expected waiting time to even the second record is infinite, both the median and the mode of the waiting times are finite. Also, it can be shown that the ratio of waiting times to r and $r + 1$ records tends to Euler number e;

$$\frac{median(W_{r+1})}{median(W_r)} \approx e = 2.718\ldots$$

even for $r = 4, 5, 6, 7, 8$. Note that "small r" does not mean "small sample size." Table below shows this ratio and the median for W_r. From this table, it is clear that fewer than eight record highs are expected in a sample size of $n = 1000$. The expected value of W_r is infinite even for $r = 2$.

Ratio of Medians of Waiting Times

Record Number r	2	3	4	5	6	7	8
Median W_r	4	10	26	69	183	490	1316
Med W_r /Med W_{r-1}		2.50	2.60	2.65	2.65	2.68	2.69

Using the table, after seeing the second record, we should wait, on average (in the median sense), ten observations to see the third record.

This suggests a geometric increase with rate of 2.718. . . If we assume that one unit of attempts was needed to arrive at the second record, the total number of attempts to arrive at record number 20 may be calculated as

$$1 + e + e^2 + \cdots . + e^{18} = 103{,}872{,}541.$$

This is a slight overestimation, as for early records, the ratios are less than e. This leads to probability estimates for a new record for one-hundred-meter dash as

0.152461 for the next year 0.562681 for the next 5 years 0.808753 for the next 10 years

Recall that after seeing the second record, the median wait time to the third record is ten observations (attempts). Other results regarding W_r include a law of large numbers, $log\, W_r /r \to 1$, and a result indicating that $log\, W_r$ is approximately equivalent to the arrival time sequence of a Poisson process. Since sports records are more frequent than records generated by independent and identically distributed sequences, it is possible to model $log\, W_r$ as a nonhomogeneous Poisson process.

Let us look at the data for long jump. For long jump, the fifth record was set in 1991. Using the theory of records, seventy-three attempts is needed to produce five records, and these should have occurred during the period 1962–1991 (thirty years). This leads to geometric increase with rate i = 1.055. Noting that the waiting time to the sixth record is 183 attempts, it takes (in median sense) forty-nine years for a new record to be set. This means waiting till the year 2040. The return period of the present record (8.95) was found to be 64.5 years based on the tail model I obtained.

We end this part by noting that, rather than records and waiting times between them, one could consider improvements of equal size and analyze the corresponding waiting times. This seems a reasonable approach, since as records improve, increase in number of attempts could offset the decrease in number of record-breaking performances. For example, consider the rise in pole vault records and their waiting times shown in the following table.

Data for Pole Vault

Improvement (feet)	Number of Years
14 to 15	13
15 to 16	22
16 to 17	1.5
17 to 18	7
18 to 19	10
19 to 20	10

Here, one can consider smaller improvements and apply some of the classical statistical methods. In the case of pole vault, for example, the goal of such analysis should be to predict the number of years it would take to go from twenty to twenty-one.

Prediction of Future Records

Apart from intrinsic interest, there are several medical and physiological reasons why we would like to know how fast a human being could, for example, run a short distance such as one-hundred-meter dash or a medium distance such as four hundred meters. While there is a general agreement among the physiologists and physical educators about existence of, for example, an upper limit for such a speed (or a lower limit for the time needed to run this distance), the limit is not known at the present time. Because of a great interest in this question, apart from physiological research, there have been some attempts to estimate (predict) the limits via mathematical modeling. Recall that the performance of a human being in athletic events is an issue of great popular interest.

Methods

Prediction of the future records can be divided into short term and long term. Different approaches can be taken depending on goal of study and the data one may use.

1. Using all the available data, one may use general trend analysis.
2. Considering the fact that both records and improvements are decreasing, one may use ideas from population biology and apply models such as logistic that has an asymptote, which makes its tail to level off.
3. Data analysis and modeling may be performed using only the best record of, for example, each year. The modeling may then be carried out using the extreme value theory. Recall that there are three types of extreme value distribution of which one has a lower bound to represent the ultimate record.
4. Data analysis and modeling may be performed using only the times below a chosen time such as ten seconds. The modeling may then be carried out using the threshold theory. This theory models data on the lower tail of the distribution for times. Recall that, here, too, there are three types of tail models of which one has a lower bound representing a short tail model and the ultimate record.
5. Data analysis and modeling may be performed using only the records. The modeling may then be carried out using the theory of records. Recall that this theory models number of records, time of the records, time interval between records, and, finally, size of the records.

   Since sport records occur more frequent than what this theory predicts, some modification to account for factors that lead to this is needed. Here, one can use population growth or growth in population of participants for this adjustment.

Trend Analysis

Trend analysis seeks to decompose the data into a systematic (deterministic, trend, signal, talent) part and a random (stochastic, noise, chance) part. When the form of the systematic part is known, it is easy to separate that from the random part. In the absence of information pertaining to the either systematic or random parts, smoothing is used for separation. This is usually carried out assuming that the systematic part is smooth and the random part is rough. Smoothing is an exploratory operation, a means of gaining insight into the nature of data without precisely formulated models or hypotheses. Smoothing is often achieved by some sort of averaging (low-pass filtering). Once the smooth part is determined, the difference between the original message and the smooth part is used to present the rough part. The rough part is usually utilized to make reliability statement regarding the systematic part. For data with a time index (time series), one popular smoothing technique is the so-called moving averages. The idea is to average the neighboring values and to move it along the time axes.

The idea is to smooth the data using techniques such as moving average in stock market to recover the trend OR by replacing the data points by a smooth curve using methods such as regression analysis. When smoothing, we can weight more recent performances (times) or records more than other. Finally, we can assume that improvements are proportional to the last record and use the logistic model.

General Methods

To analyze sports records, a large number of investigators have proposed models that are made up of a deterministic component

$z(t,\theta)$ to account for the trend and a stochastic component $x(t)$ to account for the variation—that is,

$$y(t) = Z(t,\theta) + x(t) \tag{1}$$

For example, we can assume that $x(t)$'s are independent and identically distributed random variables. Particular distributions considered for $x(t)$ are, normal, Gumbel, and the generalized extreme value. For $Z(t,\theta)$ the following linear, quadratic, and exponential-decay models were examined:

$$Z(t,\theta) = \theta_0 - \theta_1 t, \, \theta_1 > 0 \tag{2}$$

$$= \theta_0 - \theta_1 t + \theta_2 t^2 / 2, \, \theta_1 > 0 \tag{3}$$

$$= \theta_0 - \theta_1 [1 - (1 - \theta_2)^t] / \theta_2, \, \theta_1 > 0, \qquad 0 < \theta_2 < 1 \tag{4}$$

Using the maximum likelihood method and numerical approximation procedures, these models were applied to some data. The normal distribution was found to be the most appropriate among the three distributions, although it was noted that the choice of distribution was not crucial for forecasting purposes. It was also noted that the quadratic or exponential model did not provide a significant improvement over the linear model.

When estimating the limiting time (ultimate records) from exponential-decay model, the standard errors were so large that made the predictions meaningless. Smith has implied that the choice of model may be inappropriate and acknowledged the wide variability of estimates corresponding to different error distributions and different portions of the series. In fact, he is doubtful that the use of such methods, in general, and model (4), in particular, can produce meaningful performance estimates for the distant future.

Although useful in some cases, additive models of the form (1) may not be appropriate for sports data as they imply no relation between variation in $x(t)$ and change in $z(t,\theta)$. In fact, it is reasonable to expect decrease in variability in the latter portion of the data because performances are getting closer to the ultimate record and significant improvements are less likely now than, say, fifty years ago. Also, since most world-class runners remain competitive for a number of years (usually between three and six), we may see some dependency between neighboring performance measures.

Noubary (1994) has suggested the following models:

$$\log y(t) = \theta_0 - \theta_1 t + x(t), \qquad \theta_1 > 0 \qquad (5)$$
$$\log y(t) = \theta_0 - \theta_1 t + \theta_2 \log t + x(t), \qquad \theta_1 > 0 \qquad (6)$$

where $\{x(t), t = 1, 2, \ldots\}$ is a zero-mean stationary process. Note that models (5) and (6) can alternatively be written in multiplicative form as

$$y(t) = \theta_0^* e^{-\theta_1 t} x^*(t), \qquad \theta_1 > 0 \qquad (7)$$
$$y(t) = \theta_0^* t^{\theta_2} e^{-\theta_1 t} x^*(t), \qquad \theta_1 > 0 \qquad (8)$$

where $\theta_0 = \log \theta_0^*$ and $x(t) = \log x^*(t)$. These models imply that both means and variances vary with time—that is, unlike the additive model where variation in $y(t)$ is independent of t, here, the variation decreases as t increases. As a result, the standard errors of the future records are smaller than those of an additive model, and therefore the likelihood of obtaining a meaningful prediction is higher.

Application of Theory of Records

Consider men's one-hundred-meter data. Recall that theory of records attempts to find answers to the following questions:

1. How many records do we expect to observe in n attempts? The table and the graph below present the expected number of records for $n \leq 85$.

Number of Attempts (Years) (n)	Expected Number of Records (R_n)
3	1.83
6	2.45
9	2.83
12	3.10
15	3.32
18	3.50
21	3.65
24	3.78
27	3.89
30	3.99
33	4.08
36	4.17
39	4.25
43	4.35
50	4.50
60	4.67
70	4.80
80	4.96
81	4.97
82	4.99
83	5.00
84	5.01
85	5.03

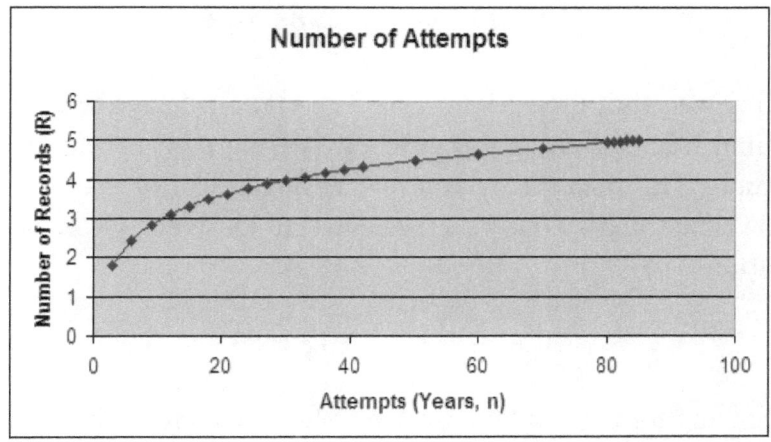

2. How long will the present record stand?
3. What is the probability of observing a new record?
4. Can we predict future records? If yes, what would be the value of the next and the future records?
5. Is there an ultimate record? If yes, how do we estimate that?

Using the theory, we can only provide answer only if the data were generated by a sequence of independent and identically distributed random variables. Sport records are usually more frequent than what the theory predicts partly because of

1. improvement in training, coaching, equipment, diet, etc.;
2. increase in participation or attempts;
3. participation of same runners, etc.

To account for this, some adjustment is therefore necessary. I have treated the problem as if either participation has increased with time or more competitions have taken place so that the chance of setting a new record was increased. From the above table, for independent and identically distributed observations with five records, we need eighty-three attempts (year), since

$$1 + 1/2 + \ldots + 1/83 = 5.$$

Suppose that data contains five records in forty-three years. Assume that i is the geometric rate of increase in number of attempts. This means that every year, the number of attempts has been i time the previous year. We can find its value by solving the equation:

$$1 + i + i^2 + \ldots + i^{42} = 83.$$

Solving for i, we obtain a value of 1.027, which is a 2.7 percent rate of increase (2.7 percent more attempts per year).

This means that attempts are increased as

1	2	3	...	43
1	1.027	$(1.027)^2$		$(1.027)^{42}$

Now, to predict records for the next ten years, we set

$$n_1 = 83, \ n_2 = (1.027)^{43} + \ldots + (1.027)^{52} = 36.51.$$

For this example, using result (a), the probability of a new record in next ten years is

$$n_1 / (n_1 + n_2) = 0.31.$$

Using the maximum likelihood method, a better estimate is $n_1 = 73$. This will lead to a smaller i value.

This approach seems reasonable as it could account for other factors such as advanced training programs, better equipment, diet, coaching, and even use of steroids–all of which increase the chance of setting new records. To clarify, suppose that the probability of breaking the record is q in n attempts. Suppose further that we could have had more participation increasing the number of attempts to 3n.

Then the probability that at least one group break the record is

$$1- (1-q)^3.$$

For example, this probability is 0.143 for $p = 0.05$ and 0.271 for $p = 0.10$.

Boston Marathon

This is an example of participation increase. Using regression, the model given below was found for the number of participants since 1970 as a function of the year ($R^2 = 0.938$).

Number of participants in year $t = -1294 + 1088t - 57.5t^2 + 1.25t^3$

The year 1970 is selected as the starting year because during this year, a qualifying time was introduced for participation.

We can also use world's population growth. There are already models for this. Berry (2002) has discussed the effect of population increase on the breaking of sports records and has introduced the following exponential model for the growth of the world's male population since 1900.

Population in year $t = 1.6 \exp [0.0088(t - 1900)]$

Note that this means a geometric increase of $\exp(0.0088) - 1 = 0.00884$ per year since year 1900. We can use this assuming that the number of participants or attempts is proportional to the population size at time t. However, this approach does not use information from the sports itself and the way records were set. In other words, it is the same for all sports regardless.

Records and Order Statistics

This section describes a method based on distributional

structure of order statistics. It accounts for dependency between the records and the facts that records have improved with time and the amount of improvements have become smaller.

Several studies performing trend analysis have concluded that the simple linear trend performs as good as any other more-complicated trend functions examined. In fact, many investigators have shown that records have almost a linear trend. Considering this, a method is developed based on modeling the records using distribution of consecutive order statistics from a selected distribution. To clarify, consider the data for one-hundred-or four-hundred-meter runs. Assume that after a suitable scaling distribution of time for the year one (first year) is the same as that for (n+1)st order statistics X_{n+1} from a sample of size $2n + 1$ taken from a selected distribution. Suppose that the successive times have distributions same as $X_r, r = n, n - 1, ...$ This implies that future times have both smaller means and smaller variances. Additionally, it assumes that the times are correlated. For sports such as long jump or when speed is used in place of time, one should consider $X_r, r = n + 2, n + 3, ...$ Here, increases in r results in increase in means but decrease in variances. Note that the variances decrease as we move away from the middle order statistics in both directions. Note also that since the probability density functions of consecutive order statistics overlap, this method can be applied to the annual best times where the best time of the year j may or may not be better than the best time of the year $j - 1$. Additionally, here, the correlation structure of order statistics can be used to model the correlation between the best annual times or records.

To clarify, consider the order statistics $0 \leq X_1 \leq X_2 \leq ... \leq X_r \leq ... \leq X_{n+1} \leq ... \leq X_{2n+1} \leq 1$ for a sample of size $2n + 1$ from a standard uniform distribution $u(0,1)$, where 0 and 1 represent respectively the lower and upper bounds of the attribute measured. It is known that the order statistic from a uniform distribution have beta distribution. The probability density function and cumulative distribution function for X_r are respectively

$$f_{x_r}(x) = r \binom{2n+1}{r} x^{r-1}(1-x)^{2n+1-r} \qquad 0 \le x \le 1$$

$$F_{x_r}(x) = \sum_{k-r}^{2n+1} \binom{2n+1}{k} x^k (1-x)^{2n+1-k} \qquad 0 \le x \le 1$$

The mean, variance of X_r and covariance of X_r and X_s are respectively

$$\mu_r = \frac{r}{2(n+1)}, \qquad \sigma_r^2 = \frac{r(2n+2-r)}{4(n+1)^2(2n+3)}, \qquad \sigma_{rs} = \frac{r(2n+2-s)}{4(n+1)^2(2n+3)}.$$

The expressions for mean and variance imply that for a given n, they both decrease as r decreases (or both increase as r approaches the central values). The maximum value of the variance occurs at $r = n+1$. Also, for fixed r, the variance decreases as n increases. Note that this distribution is symmetric, has a mean of 1/2 and variance of $(2n+3)/4$, and tends to normal as n increases.

For uniform distribution, the joint distribution of two consecutive order statistics is given by

$$f_{r,r+1}(x_1, x_2) = \frac{(2n+1)! \, x_1^{r-1}(1-x_2)^{2n-r}}{(r-1)!(2n-r)!} \, , \qquad x_1 \le x_2$$

This is useful for analysis of two consecutive times. It is interesting to note that the joint distribution of all order statistics depends only on the values of the first and last measurements.

The pdf of X_r given $X_s = x_2$ $(r < s)$ is given by

$$f_{x_r|x_s}(x_1|x_2) = \frac{(s-1)! \, x_1^{r-1}(x_2-x_1)^{s-r-1} x_2^{1-s}}{(r-1)!(s-r-1)!} \, , \qquad x_2 \ge x_1$$

This is useful for analysis of future times given a past time. Also, the pdf of X_s given $X_r = x_1$ is

$$f_{x_s|x_r}(x_2|x_1) = \frac{(2n+1-r)!\,(1-x_2)^{2n+1-s}(x_2-x_1)^{s-r-1}(1-x_1)^{r-2n+1}}{(2n+1-s)!\,(s-r-1)!}$$

This is useful for analysis of past times given a future time. For two consecutive order statistics, we have

$$F_{X_r|X_{r+1}}(x_1|x_2) = (x_1/x_2)^r \qquad\qquad x_1 \le x_2$$

$$F_{x_{r+1}|x_r}(x_2|x_1) = 1 - \left(\frac{1-x_2}{1-x_1}\right)^{2n+1-r} \qquad\qquad x_2 \ge x_1$$

This is useful for analysis of the next time given the present time.

Let $X_r = U$ and $X_s - X_r = V$ ($r < s$), then the distribution of the difference of two order statistics is

$$h(v) = \frac{(2n+1)!}{(r-1)!\,(s-r-1)!\,(2n+1-s)!}\, v^{s-r-1} \int_0^1 u^{r-1}\,(1-u+v)^{2n+1-s}\,du$$

This can be used to analyze the improvements.

Normalizing the range

Competition in sports such as one-hundred-meter runs takes place several times a year. We can choose a specific year as a starting point, find the mean and standard deviation of winning times, and calculate $\bar{x} \pm 2s$. We then use $\bar{x} - 2s$ as 0 and $\bar{x} + 2s$ as 1.

Alternatively, we can look at the records set so far and treat them as order statistics from a uniform distribution. For example, consider records of one-hundred-meter runs prior to the last record given below. There are total of nineteen records set. Considering this as an order statistics from a uniform distribution, we can use the first record as 1 and use last record- (last record-first record)/18 as 0.

For above data, 1 represents 10.40 and 0 represents

9.69-(10.40-9.69)/18 = 9.65. Thus, the distribution of the next record is the same as the distribution of the first order statistics for a sample of size of forty-one. Note that our unit here is 10.40-9.65=0.75. The pdf, mean, and variance of the next record are respectively

$$f(x) = 41(1-x)^{40} \quad 0 \le x \le 1,$$

9.65- 1/42(0.75) = 9.63, and (41/(4(20)2(43))(0.75) = 0.000447.

Here, 1 represents 9.69 and 0 represents 9.63-(10.40-9.63)/19 =9.61, and our unit, as before, is 0.75. For example, the probability that the next record is better than 9.67 is $1 - F(0.02/0.75) = 1 - (1 - 2/75)^{41} = 0.67$. As we know, the new record is 9.58. The probability that the next record is better than 9.58 is $1 - F(0.11/0.75) = (1 - 11/75)^{41} = 0.0015$. This again indicates the greatness of the Bolt.

It is interesting to note that since for standard uniform distribution the difference between successive order statistics is equal to the area under the pdf between them, we could estimate n by first calculating the average difference and setting that equal to $1/(n+1)$. For above data, the estimated value of n is 17.

Next, consider all records of one-hundred-meter runs including the last one, 9.58. There are total of twenty records set so far. Again, we consider this as an order statistics from a uniform distribution with the first record as 1 and the last record- (last record-first record)/19 as 0.

Here, 1 represents 10.40 and 0 represents 9.58-(10.40-9.58)/19 = 9.537. Thus, the distribution of the next record is the same as the distribution of the first order statistics for a sample of size of forty-three. Note that our unit here is 10.40-9.537 = 0.863. The pdf, mean, and variance of the next record are respectively

$$f(x) = 43(1-x)^{42}, \ 0 \le x \le 1, \quad 9.537- 1/44(0.863) = 9.517, \text{and}$$
$$(43/(4(21)^2(45))(0.863) = 0.0004675$$

Here, 1 represents 9.58 and 0 represents 9.517- (10.40-9.517)/20 = 9.473, and our unit, as before, is 0.863. For example, the probability that the next record is better than 9.56 is $1 - F(0.02/0.863) = (1 - 2/86.3)^{43} = 0.3649$. The probability that the next record is better than 9.40 is $1 - F(0.18/0.863) = (1 - 18/86.3)^{43} = 0.0000428$. It is interesting to note that Bolt believes that the world record will stop at 9.40.

We can also determine the time for a specific probability. For example, if $(1 - x/0.863)^{43} = .01$, we get $x = 9.49$.

Next, let us consider all the times below ten seconds. There are total of twenty-two data points. Considering this as an order statistics from a uniform distribution, we can use the first record as 1 and use last record- (last record-first record)/21 as 0.

Here, 1 represents 9.95 and 0 represents 9.58-(9.95-9.58)/21 = 9.562. Thus the distribution of the next record is the same as the distribution of the first order statistics for a sample of size of forty-seven. Note that our unit here is 9.95- 9.562= 0.388. The pdf and mean of the next record are respectively

$$f(x) = 47(1 - x)^{46}, \ 0 \le x \le 1,$$

$$9.562 - 1/48(0.388) = 9.554$$

Here, 1 represents 9.58 and 0 represents 9.554- (9.95-9.554)/22 = 9.536, and our unit, as before, is 0.388. For example, the probability that the next record is better than 9.56 is $1 - F(0.02/0.388) = (1 - 2/38.8)^{47} = 0.083$. The probability that the next record is better than 9.40 is $1 - F(0.18/0.388) = (1 - 18/38.8)^{47} = 0.000000000000188$. It is interesting to note that Bolt believes that the world record will stop at 9.40.

We can also determine the time for a specific probability. For example, if $(1 - x/0.388)^{47} = 0.01$, we get $x = 9.544$.

Summary

Excluding Bolt's last record, 9.58, the probability of a record better than 9.58 is 0.0015. This again indicates the greatness of the Bolt.

Including Bolt's last record, 9.58, the probability of a record better than 9.56 is 0.3649. The probability of a record better than 9.40 is 0.0000428.

It is interesting to note that Bolt believes that the world record will stop at 9.40.

We can also determine the time for a specific probability. For example, if $(1 - x/0.863)^{43} = 0.01$, we get $x = 9.49$.

Considering the times below ten seconds, the probability that the next record is better than 9.56 is 0.083.

The probability of a record better than 9.40 is 0.000000000000188. Note that this probability is expected to be very small and should not be confused with the ultimate record.

We can also determine the time for a specific probability. For example, for $(1 - x/0.388)^{47} = 0.01$, we get $x = 9.544$.

NR	RESULTS	NAME	NOC	DATE
19	9.69	Usain Bolt	JAM	16/08/2008
18	9.72	Bolt	JAM	31/05/2008
17	9.74	Powell	JAM	09/09/2007
16	9.77	Powell	JAM	14/06/2005
15	9.79	Greene	USA	16/06/1999
14	9.84	Bailey	CAN	27/07/1996
13	9.85	Burrell	USA	06/07/1994
12	9.86	Carl Lewis	USA	25/08/1991
11	9.90	Burrell	USA	14/06/1991
10	9.92	Carl Lewis	USA	24/09/1988
9	9.93	Smith	USA	03/07/1983
8	9.95	Jim Hines	USA	14/10/1968
7	9.90	Jim Hines	USA	20/06/1968
6	10.00	Hary	N/A	21/06/1960

5	10.10	Williams	USA	03/08/1956
4	10.20	Owens	USA	20/06/1936
3	10.30	Williams	CAN	09/08/1930

Upon applying the estimation method discussed earlier, the maximum likelihood value of n was found to be 62. Thus, the distribution of the next record is same as X_{44} from a sample size of 125. Parameters n and i where the observed data are the order statistics from a sample of given or variable sizes.

Alternatively, we can look at the records set so far and treat them as order statistics from a uniform distribution. For example, consider records of one-hundred-meter runs given below. There are total of nineteen records set so far. Considering this as an order statistics from a uniform distribution, the expected value of the area under the pdf to the left of the present record is $1/20 = 0.05$. Using this, we can either use (the present record)$(19/20)$ as our 1 or replace $19/20$ by $38/(2x19+1) = 38/39$ and (the present record) $(38/39) = 0.944154$ as 1.

Consider a time series $X_t, t = 1, 2, ..., n$. Suppose that X_t's not identically distributed but are not in this sense very different from each other. Specifically, they have a slightly different distribution and differences such as increasing mean and variance or increasing mean and decreasing variance, etc. The analysis of time series involves studying dependency structure of observed series through examining autocorrelation and spectrum.

Suppose that the correlation structure of the observation can be modeled using a stochastic process with a probability structure similar to order statistics from an assumed distribution.

How Fast?

I have worked for years to combine my interests in mathematics, statistics, and sports. After Beijing Olympic and Berlin's competitions, I was approached by press around the world because of the work I have done on prediction of sport records and their relevance to records set in sprint events recently.

In the last two decades, the athletics performances and related issues have received considerable attention by physiologists, physical educators, and public. One aspect of interest has been the improvement of the performances over time to address the question of predicting the future performances.

Research in this area includes short-term prediction (e.g., what would be the next Olympic record and long-term prediction? What would be the ultimate record?). I have done some research in both areas and have developed methods for prediction of both. My works have some advantages over most existing methods in two directions:

(1) They include margin of error. For example, I have estimated that with a 95 percent confidence, the ultimate time for one-hundred-meter dash is 9.44 seconds.

(2) They go far beyond trend analysis and utilize recent developments such as theory of records, threshold theory, and tail modeling and their modified versions.

The remarkable performances of Usain Bolt in the two-hundred-meter and one-hundred-meter sprints during the Beijing Olympic in 2008 and Berlin competitions in 2009 have prompted many, especially United States Olympic committee, to focus on predicting records. They are very interested to know just how surprising Bolt's performances in one hundred and two hundred meters were. They have found my work interesting and useful. They are working on several projects with the aim of finding (among others) an improved ability to know whether a given USA athlete (or program) is on track to be competitive against top international competitors, and if so, how is it possible to establish more meaningful and realistic goals, performance interventions, and perhaps be more objective and informed about their funding decisions? Some of these studies aim at finding answer to the following questions:

1) What is the short-term prediction (e.g., next Olympic) for the fastest time in a one-hundred-meter sprint?

2) What is the long-term prediction or ultimate record for one-hundred-meter sprint? My prediction was much lower than previously held beliefs (he believes that the ultimate time is 9.44 seconds). And last year, Usain Bolt had a record-breaking one-hundred-meter sprint with a time of 9.69 seconds during the Olympic Games in Beijing, and this year an amazing time of 9.58 seconds during the Berlin competitions.

My methods for calculating this minimum time have been of interest to the United States Olympic committee. My methods are different from most other predictions because they are accompanied with a margin of error. The United States Olympic committee is interested in a program that would compare U.S. runners with runners from other countries and how they can interfere with an athlete's performance to make him more competitive. The method I myself developed provides a lower bound for how far they can expect athletes to go. I have been interested in the tail of the distribution for data. I have used the theory of records, threshold data, and tail modeling to make my predictions.

Once the gun sounds, Usain Bolt seems to test the very limits of the race—the human race. The Jamaican sprinting sensation put on another amazing performance at the world championships on Sunday, shattering his own record in the one hundred meters by 0.11 seconds to take it down to an almost inhuman 9.58 seconds. Maybe "inhuman" is a bit too strong, but the man is certainly on another level.

Note that this result was the biggest improvement in the one-hundred-meter record since electronic timing began in 1968. Bolt's not done either. A few days later, he cruised into the semifinals of the two hundred, and he also figures to lead his nation's four-hundred-meter relay. At that, who knows how low he can go? He's certainly willing to try.

"Personally, I think I have more work to do," Bolt said after winning the one hundred meters title at the same Olympic

Stadium where Jesse Owens won four gold medals at the 1936 Berlin Games.

Several researchers have done studies recently to predict how fast a man can run one hundred meters. The latest from Tilburg University in the Netherlands predicts that someone will eventually be able to break the tape at 9.51 seconds. Bolt, who has set three records at the one-hundred-meter distance with times of 9.72, 9.69 and 9.58, is already looking to rip that theory apart.

"I said 9.4," Bolt said. "I think the world records will stop at 9.4."

"You'd have to think he can't keep getting faster, but we wouldn't put it past him," Ladbrokes spokesman Robin Hutchinson said in a statement. Bolt became the premier runner in the world when he ran 9.72 in May 2008–only two and a half months before the Olympics, where he lowered the mark to 9.69. At the Bird's Nest in Beijing, he outdid himself, easing up at the end of the one hundred, mugging for the cameras even before he showboated across the finish line–and also setting records in the two hundred and four hundred relays.

Note that one indicator of getting close to limit is that amount of improvements should decrease. This is not true both for one-hundred-meter and two-hundred-meter data.

Long Jump

Consider the data in the table that includes the best annual long jump distances corresponding to thirty-eight years 1962–1999. Note that although for small n results may vary for different choices of m, they all converge to similar values as n increases. When n is not very large, we may choose m as follows. Since both $m \to \infty$ and $m/n \to \infty$ are necessary to provide accurate estimates, a function with the following property provides a good choice,

$$m(n) = O(n^{1/2})$$

as it guarantees the same rate of convergence for both requirements. However, although this is a reasonable choice, there are as good or even "better" choices. This is important noting that the function $m(n)$ plays a critical role (Smith 1987; Coles 2001) in studying the tail of a distribution.

For example, we may choose $m(n)$ that includes information regarding the record times (the times when records were set) and the inter-record times (the time between successive records). These two sequences are particularly relevant, and their influence on prediction of future records is clear.

Long Jump Distances (values with * are records)

Year	Distance (m)	Year	Distance (m)	Year	Distance (m)
1962	8.31*	1975	8.45	1988	8.76
1963	8.30	1976	8.35	1989	8.70
1964	8.34*	1978	8.32	1990	8.66
1965	8.35*	1979	8.52	1991	8.95*
1966	8.33	1980	8.54	1992	8.58
1967	8.35	1981	8.62	1993	8.70
1968	8.90*	1982	8.76	1994	8.74
1969	8.34	1983	8.79	1995	8.71
1970	8.35	1984	8.71	1996	8.58
1971	8.34	1985	8.62	1997	8.63
1972	8.34	1986	8.61	1998	8.60
1974	8.30	1987	8.86	1999	8.60

Assume that the data contains r records. Let T_r denote the time between the last and penultimate records and t_r denote the time the last record has held to date. Then it can be shown that the following choice proposed by Tata (1986) satisfies the required conditions:

$$m(n) = \sqrt{eT_r} + \sqrt{t_r} = \sqrt{2.718282 T_r} + \sqrt{t_r}$$

Comparison using simulated data on beta distribution (unfavorable cases) shows that this is a better choice compared to choices such as $m(n) = O(n^{1/2})$. Further, the results very much depend on the last value of T_r as expected. In fact, the time between the two latest records being not inordinately large (or small) is fairly essential for the reasonably accurate estimate. This applies to choices such as $m(n) = O(n^{1/2})$ as well, although it does not depend on T_r.

For the long jump data, n is not large, and therefore we do not expect to obtain a very accurate estimate for the probabilities of future performances. However, we use this data for demonstration. For this data,

$$t_r = 1999 - 1991 = 8, \quad T_r = 1991 - 1968 = 23.$$

Using these, we get

$$m = \sqrt{23} + \sqrt{8} = 10.74 \approx 10$$

and

$$a^* = \frac{1}{10} \sum_{i=1}^{10} \ln x_i - \ln x_{11} = 2.173344 - 2.163323 = 0.01$$

This leads to the following model for the tail $m(n)$

$$\bar{F}(x) = 10/38 \, (x/8.70)^{-100}$$

$$\bar{F}(x) = 10/38 (x/8.70)^{-100}$$

From this model, the values of $P(X > 8.95)$ and $P(X > 9.00)$ are respectively 0.0155 and 0.00887 for one year and $1 - (1 - 0.0155)^{10}$ $= 0.1446$ and 0.0852 for ten years. Also, the return period of X to exceed 8.95 is $1/0.0155 = 64.5$ years. To see whether this makes sense, consider the second-best performance (distance) 8.90. From the above model, we have

$$P(X > 8.90) = 0.0271.$$

Thus, the return period for a distance greater than 8.90 is about thirty-nine years. Data in table 1 indicates that this record was set in year 1968, and during the last forty-one years, it was exceeded only once. This agrees with the observation and provides support for the tail model obtained.

We end this part by noting that the probability calculations may significantly change if new records are set or current records survive for a long time. For example, for long jump, no record was set during 2000–2004. The best distances for this period are respectively 8.65, 8.41, 8.52, 8.53, and 8.60–all smaller than even the best eleven distances of the past thirty-eight years prior to the year 2000. Including these data and repeating the calculations, the probability for setting a new record $P(X > 8.95)$ are now 0.02318 for one year and 0.20906 for ten years, respectively.

A practice problem. As we know, the record of this event was set in 1991 and surprisingly still holds.

1. Find the complete data and repeat the analysis.
2. What is the reason for this record to stay this long? Could it be because human beings are close to the limit of their ability?

Four-Hundred-Meter Run

Finally, we note that the proposed method would work better for a large data set. For instance, for men's four-hundred-meter run, the fastest times were recorded every year since 1860. The last three records, 43.18, 43.29, and 43.80, were set in years 2000, 1998, and 1968, respectively. Using this information, we get $m = 8$ and the following tail (lower tail) model for $x < 44.40$:

$$\bar{F}(x) = 8/141(44.40/x)^{-90.91}$$

From this, the values of P(X < 43.10) and P(X < 43) are respectively 0.0038 and 0.0031 for one year and 0.0374 and 0.03057 for ten years.

One-Hundred-Meter Run

Consider the data for men's one-hundred-meter run for the period January 1, 1977, to September 1, 2009, after IAAF required fully automatic timing to the hundredth of the second. For this, there are between thirty and forty data points some being invalidated because of drug use and some being discarded (including Gay's 9.68 time on June 2008 during the 2008 U.S. Olympic Trials) because of the wind speed exceeding the IAAF legal limit. Here, the time between the two latest records and the time the last record has stood to date are inordinately small. So rather than using $m(n)$ based on this information, here, we use $\sqrt{n}$ instead. This is a good choice noting that the rate of convergence is the same for both requirements. For the data considered, the closest integer to $\sqrt{n}$ is $(m(n) = 6)$. The top six legal times are

$$x_1 = 9.58, \; x_2 = 9.69, \; x_3 = 9.71, \; x_4 = 9.72, \; x_5 = 9.72, \; x_6 = 9.74.$$

These were set respectively by Bolt, Bolt, Gay, Powell, Bolt, and Powell. Using these for $x < x_7 = 9.77$, we get

$$\bar{F}(x) = 6/33(9.77/x)^{-126.694}$$

and P($X \le 9.5$) and P($X \le 9.55$) are respectively 0.0052 and 0.0102.

Bolt's Effect

To measure the effect Bolt has had so far and may have in the future, let us calculate the same probabilities by excluding his three records. The top six legal times are, then,

$$x_1 = 9.71, \; x_2 = 9.72, \; x_3 = 9.74, \; x_4 = 9.77, \; x_5 = 9.79, \; x_6 = 9.84.$$

These were set respectively by Gay, Powell, Powell, Powell, Greene, and Bailey. Using these for $x < x_7 = 9.84$, we get

$$\overline{F}(x) = 6/33(9.84/x)^{-124.954}$$

and P(X $\leq$ 9.5) and P(X $\leq$ 9.55) are respectively 0.00225 and 0.00433. As can be seen, these probabilities are significantly lower than the probabilities when Bolt's records were included. Also, we have P(X $\leq$ 9.58) = 0.0064, which demonstrates that Bolt is in a different league.

Usain Bolt's Two-Hundred-Meter Individual Performance

Consider Usain Bolt's individual performance in two-hundred-meter run. His record, 19.30, in the national stadium in Beijing (August 20, 2008) was considered extraordinary by many experts. As such one may ask, was his world record in the two hundred meters within or outside of the realm of what could have been expected from him? Given his past performance, how well within or outside of what could have reasonably expected was Bolt's two-hundred-meter performance in Beijing Olympic? To provide an answer, consider his performance history prior to the 2008 Olympic.

| 2007 | 19.75 | Kingston (NS), JAM | 24/06/2007 |
| 2006 | 19.88 | Lausanne | 11/07/2006 |

2005	19.99	London (CP)	22/07/2005
2004	19.93	Devonshire	11/04/2004
2003	20.13	Bridgetown	20/07/2003
2002	20.58	Kingston, JAM	18/07/2002
2001	21.73	Debrecen	14/07/2001

Since these are his best times, we may consider them as tail data. Using this information, we have $m = 3$, and the tail (lower tail) model for his performance prior to the 2008 Olympic is

$$\bar{F}(x) = 3 / 7(19.99 / x)^{-145.61}$$

From this, the values of $P(X < 19.32)$, $P(X < 19.31)$, and $P(X < 19.30)$ are respectively 0.00299, 0.00278, and 0.00257. This shows that his world record in the two hundred meters was outside of the realm of what could have been expected from him. In fact, given his performance history prior to the Olympic, one should have expected him to break his own record, 19.75, with a probability of 0.0738. This is because his best time prior to this record (19.75) was 19.88 set in 2006.

Now, consider his yet more amazing new record, 19.19 (August 20, 2009), set in Berlin. We could, following the same line, show that this record was so unexpected too. But, instead, let us include all his data and make some predictions. With nine data points, we have $m = 3$ and the following tail (lower tail) model for his performance,

$$\bar{F}(x) = 3/9(19.88/x)^{-41.96}$$

From this, the values of $P(X < 19.15)$ and $P(X < 19.10)$ are respectively 0.0694 and 0.0622.

Finally, although we did not have enough data to carry the same analysis for his one-hundred-meter runs, the same obviously applies to that event too after he recently shattered the world record. In fact, his Olympic record 9.69 and the new record 9.58 set in Berlin support the view that his performance in this event was even more than just extraordinary.

Conclusion

A procedure is used for calculation of probabilities of future athletic records based on modeling the tail of the distribution for performance measures. The results obtained for men's long jump and four-hundred-meter runs are reasonable. For long jump, the last two records 8.95 and 8.90 are significantly greater than the third best record 8.35, indicating a medium or long-tail model. Data regarding the performance of Usain Bolt have a similar characteristic, as his Olympic times are significantly lower than his performances prior to that. Additionally, the use of medium or long-tailed models avoids difficulties posed by intractable likelihood equation and the complications regarding ultimate records and their estimation.

Methods Based on Theory of Records (Revisited)

As discussed earlier, the theory of records deals with values that are strictly greater than or less than all previous values. Usually, the first value is counted as a record. Then a value is a record (upper record or record high) if it is superior to all previous values.

The study of record values, their frequencies, times of their occurrences, their distances from each other, etc., constitutes the theory of records. Formally, the theory deals with four main random variables:

(1) the number of records in a sequence of n observations;
(2) the record times (indices);
(3) the waiting time between the records; and

(4) the record values.

The theory of records has not yet fully exploited for addressing questions regarding the prediction of sports records. This is mainly because the results of theory of records for independent and identically distributed sequences are not directly applicable to sports. In fact, in most sports, records occur more frequently than what the theory predicts. To account for this, one may treat the problem as if participation has increased with time or more attempts are made so that the probability of setting a new record was increased (Noubary 2005).

To predict the future records, Noubary (2005) has developed a simple method utilizing the following results of the theory of records for independent and identically distributed sequence of observations.

(a) If there is an initial sequence of n_1 observations and a batch of n_2 future observations, then the probability that this additional batch contains a new record is $n_2/(n_1+n_2)$.
(b) For large n, $P_{r,n}$ the probability that a series of length n contains exactly r records is given by ($\gamma = 0.5772$).

$$P_{r,n} \sim \frac{1}{(r-1)!n}(\ln(n)+\gamma)^{r-1}$$

To demonstrate the method, let us consider the one-hundred-meter data for the period 1912–2009. The number of records set in this period is $r=20$. For this, data application of the maximum likelihood method yields n = 317,884,920 attempts. This is found by maximizing $P_{r,n}$ with respect to n. It estimates the number of independent and identically distributed attempts required to produce twenty records.

Now, suppose, for example, that i is the geometric rate of increase in attempts increase. This means that the number of attempts in any given year is assumed to be i times the number of attempts in the year before. The value of i can be found by solving the equation

$$1 + i + i^2 + \cdots + i^{97} = 317884920$$

where i^j represents the number of attempts in year j. Here, we get $i = 1.2013$, which means 20.13 percent more attempts per year.

To predict records for the future one and five years (for now, year 2010 and the period 2010–2015), we replace $n_1 = 317884920$ and $n_2 = (1.2013)^{98} = 63950619$ for the next year, and $n_1 = 317884920$ and $n_2 = (1.2013)^{98} + \ldots + (1.2013)^{102} = 477112800$ for the next five years in result (a) above. This leads to probability estimates of 0.1926 and 0.6001 for a new record to be set in the next one and five years, respectively.

Summary: Probability estimates for a new record in the next one and five years respectively as 1926 and 0. 6001.

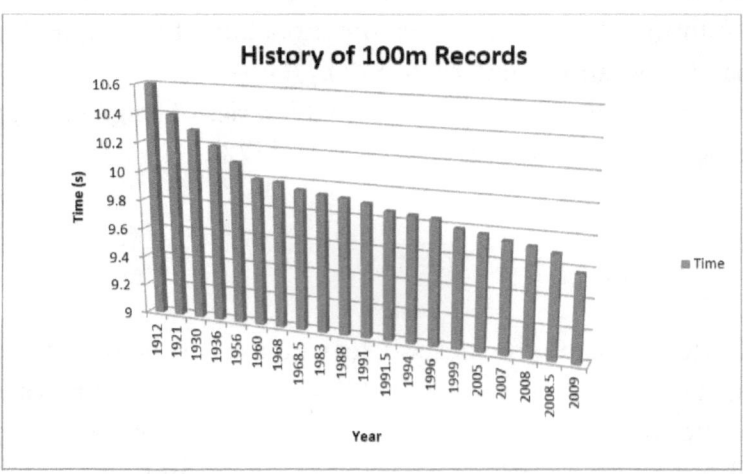

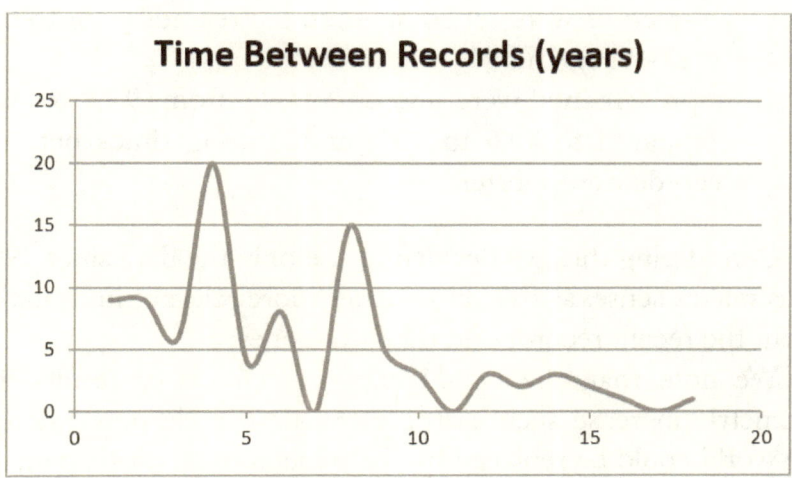

Unlike the theoretical expectation, the waiting times between the records have decreased significantly with time especially in the near past.

To account for these and other contributing factors, here we will treat the problem as an independent and identically distributed one but make up for the increase in probability and frequency of the records by inflating the number of attempts. This, as we shall demonstrate, is a reasonable approach.

Recall that Berry (2002) has used the increase in male population of the world as an adjusting factor for participation. His model for the growth of the world's male population–namely,

Population in Year $t = 1.6\ exp\ [0.0088(t-1900)]$

implies a geometric increase with annual rate of $exp(0.0088) = 1.0088388$, which for one-hundred-meter data is clearly unrealistic. When discretized, this equation implies

$$y_{n+1} - y_n = 1.0088\ y_n$$

Also, looking at the data, we see a change in rate of records and their waiting times. In fact, since1990,

1. ten records were set compare to ten records prior to that (1912–1990); and
2. improvements were more notable from 9.92 to 9.58 compared to 10.6 to 9.92, and waiting times between records were shorter.

Considering this, we decided to use only the data since 1990. This makes sense, as this data carries more relevant information about the recent records and the future ones.

We note that one would expect even better results if a geometric increase such as the increase in male population of the world could be replaced by the population of participants or by the number of attempts. So to improve the estimates, rather than geometric increase, we can, following a general approach for modeling population increase, consider models such as logistic or Gompertz or more generally a model of the form

$$y_{n+1} - y_n = H(y_n) = i * f(y_n)(1 - g(y_n))$$

where y_n denotes the number of participants or number of attempts at year n (generation n). One of the simplest and frequently used models that contain a formulation that avoids indefinite growth and represents effects of overcrowding is when i is a linear function of the last year's participation. This choice of i leads to a model of type

$$y_{n+1} - y_n = i^* y_n (1 - y_n / h) = H(y_n)$$

known as logistic equation. Here, i^* represents the rate of growth and h represents the carrying capacity. For one hundred meters, h may be the maximum number of individuals who qualify to participate in an event such as the Olympics. This type of model is reasonable for sports where usually rapid initial improvements are followed by much slower advances.

In an article in 2005, I considered the following simpler model instead that exhibits the same behavior as the logistic equation:

$$y_{n+1} - y_n = H(y_n) = i * f(y_n)(1 - g(y_n))$$
$$y_{n+1} = y_n exp[r * (1 - y_n/h)]$$

For example, the number of attempts in the future one and ten years are 412 and 4876 using $y_0 = 100$, $r^* = 0.04$ and $h = 50$. The corresponding numbers using the logistic equation are 402 and 4742. These numbers result in smaller probability estimates compared to the geometric increase of 4 percent.

Long-Term Prediction, Ultimate Record

Let

$$Y_1 < Y_2 < ... < Y_n$$

be the order of statistics of times in the men's one-hundred-meter run. Then it can be shown that a level $1 - p$ confidence interval for R_u (Ultimate Record) is given by

$$\{Y_1 - (Y_2 - Y_1) / [(1 - p)^{-\alpha} - 1], Y_1\}$$

where $\alpha = \log k(n) / \log [(Y_{k(n)} - Y_3)/ (Y_3 - Y_2)]$ and $k(n)$ is any sequence such that $k(n) \to \infty$ and $k(n)/n \to 0$. One natural choice is $k(n) = \sqrt{n}$. It can be shown that a better choice is

$$k(n) = (eT_r)^{1/2} + (t_r)^{1/2}$$

where T_r is the time between the last and previous ultimate record, and t_r is the time the last record has held to date.

Prediction of ultimate records can be carried out using, for example, a model with an exponentially-decay trend. It is also possible to consider simpler trends such as

$$x = (b + ct)/(1 + t) \text{ or } x = (a + bt + ct^2)/(1 + t + t^2)$$

and use the fact that as $t \to \infty$, $x \to c$. However, as pointed out earlier, the application of such models usually results in predictions with large standard errors, which are not useful or even acceptable (Smith 1988). This section describes a method based on tail modeling using a certain number of exceptional performances. This approach avoids the problem mentioned and provides a confidence interval for the ultimate record based on the most recent performances or records.

Let X and $X_1, X_2, ..., X_n$ represent respectively a performance measure and a set of observed values for a given sport such as one-hundred-meter run, where

$$X_1 \leq X_2 \leq \cdots \leq X_n$$

$$X_1 \leq X_2 \leq ... \leq X_n$$

Assuming that the distribution function $F(y)$ has a lower endpoint and certain conditions are satisfied, it is shown that a level $(1 - p)$ confidence interval for the minimum of Y is

$$\{X_1 - (X_2 - X_1)/[(1-p)^{-k} - 1], \ X_1\}$$

De Haan (1981) has also shown that the statistics

$$\frac{\ln m(n)}{\ln[(x_{m(n)} - x_3)/(x_3 - x_2)]}$$

is a good estimate for k.

To demonstrate, suppose we wish to estimate the ultimate

record for the one-hundred-meter run. We consider the data for the period January 1, 1977 to September 1, as in section 3.1. Recall that we need to choose an integer $m(n)$ depending on n such that $m(n) \to \infty$ and $m(n)/n \to 0$ as $n \to \infty$. Like in section 3.1, we choose $\sqrt{n}$ as it has the same rate of convergence for both requirements. For this data, the closest integer to $\sqrt{n}$ is 6 ($m(n) = 6$) and the top six legal performances are

$$x_1 = 9.58, \; x_2 = 9.69, \; x_3 = 9.71, \; x_4 = 9.72, \; x_5 = 9.72, \; x_6 = 9.74.$$

Using the above formula, we get $k = 4.419$, and a 90 percent one-sided prediction interval for the ultimate record is

$$(9.40, 9.58).$$

Bolt's effect

Now to measure the effect Bolt has had so far and may have in the future, let us find the prediction interval by excluding his three records. The top six legal performances are, then,

$$x_1 = 9.71, \; x_2 = 9.72, \; x_3 = 9.74, \; x_4 = 9.77, \; x_5 = 9.79, \; x_6 = 9.84.$$

These were set respectively by Gay, Powell, Powell, Powell, Greene, and Bailey. Using the formula, we get $k = 1.113$, and a 90 percent one-sided prediction interval for the ultimate record is

$$(9.62, 9.71).$$

The lower bound is greater than the present record set by Bolt. In other words, Bolt has already exceeded what would have been predicted to be the limit of human performance. This shows how remarkable Bolt's performance has been and how this one runner could literally change our perception of human capabilities.

Summary

A 90 percent one-sided prediction interval for the ultimate record including Bolt's three records is

$$(9.40, 9.58).$$

A 90 percent one-sided prediction interval for the ultimate record excluding Bolt's three records is

$$(9.62, 9.71).$$

1	9.58	Usain Bolt	JAM	16/08/2008
2	9.69	Usain Bolt	JAM	16/08/2008
3	9.71	Gay	USA	16/08/2009
4	9.72	Usain Bolt	JAM	31/05/2008
5	9.72	Powel	JAM	02/09/2008
6	9.74	Powel	JAM	09/09/2007
7	9.77	Powel	JAM	14/06/2005
8	9.79	Greene	USA	16/06/1999
9	9.84	Dailey	CAN	27/07/1996
10	9.84	Surin	CAN	22/08/1999
11	9.85	Burrell	USA	06/07/1994
12	9.85	Gatlin	USA	22/07/2004
13	9.85	Fasuba	Nigeria	12/05/2006
14	9.86	Garl Lewis	USA	25/08/1991
15	9.86	Frankie Fredericks	Namibia	03/07/1996
16	9.86	Ato Boldon	Trinidad and Tobago	19/04/1998
17	9.86	Francis Obikwelu	Portugal	22/08/2004
18	9.90	Burrell	USA	14/06/1991
19	9.92	Carl Lewis	USA	24/09/1988
20	9.93	Smith	USA	03/07/1983
21	9.95	Jim Hines	USA	14/10/1968
22	9.90	Jim Hines	USA	20/06/1968
23	10.00	Hary	N/A	21/06/1960

24	10.10	Williams	USA	03/08/1956
25	10.20	Owens	USA	20/06/1936
26	10.30	Williams	CAN	09/08/1930
27	10.40	Paddock	USA	23/04/1921
28	10.60	Lippincott	USA	06/07/1912

Alternatively, we can look at the records set so far and treat them as order statistics from a uniform distribution. For example, consider records of one-hundred-meter runs given below. There are total of nineteen records set so far. Considering this as an order statistics from a uniform distribution, the expected value of the area under the pdf to the left of the present record is $1/20 = 0.05$. Using this, we can either use (the present record)$(19/20)$ as our 1 or replace $19/20$ by $38/(2\times19+1)=38/39$ and (the present record) $(38/39)=0.944154$ as 1.

Consider a time series X_t, $t = 1, 2, ..., n$. Suppose that X_t's are not identically distributed but are not in this sense very different from each other. Specifically, they have a slightly different distribution and differences such as increasing mean and variance or increasing mean and decreasing variance, etc. The analysis of time series involves studying dependency structure of observed series through examining autocorrelation and spectrum.

Suppose that the correlation structure of the observation can be modeled using a stochastic process with a probability structure similar to order statistics from an assumed distribution.

Bolt and One-Hundred-Meter Records

Consider the records for men's one-hundred-meter runs. For this event, twenty records are set since 1912. The last three records, 9.58, 9.69, and 9.72, were set in years 2009, 2008, and 2008, respectively–all by Bolt. Using this information, we get $m = 2$ and the following tail (lower tail) model,

$$\bar{F}(x) = 1/10(9.72/x)^{-113.64}$$

From this, the values of P(X < 9.50) and P(X < 9.55) are respectively 0.0074 and 0.0135. Also excluding the last record (9.58), the lower tail model is

$$\bar{F}(x) = 1/19(9.72/x)^{-323.5}$$

From this, the value of P(X < 9.58) = 0.00048, which indicates how extraordinary the new record is.

Since in this example T_r and t_r are respectively 1 and zero, one may prefer to work with $m(n) = [n^{1/2}]$ instead. Here, n = 20 and $m(n) = 4$ or 5. Using these, we get

m = 4 $\bar{F}(x) = 4/20(9.77/x)^{-110.9}$ P(X < 9.50) = 0.00894 and P(X < 9.55) = 0.016

m = 5 $\bar{F}(x) = 5/20(9.79/x)^{-108.013}$ P(X < 9.50) = 0.009714 and P(X < 9.55) = 0.017

Also excluding the last record (9.58), the lower tail model is

$$\bar{F}(x) = 4/19(9.79/x)^{-162.574}$$
$$\bar{F}(x) = 5/19(9.84/x)^{-100}$$

From this, the values of P(X < 9.58) are respectively 0.0062 and 0.018, which again indicates how extraordinary the new record is.

How Long Records Last?

In this section, survival probability of sport records is investigated assuming that the number of attempts to break follows a nonhomogeneous Poisson process. Explicit formulae are derived for two practically important cases and are applied to data from the Boston Marathon.

As discussed earlier, Berry (2002) proposed and used the male population of the world as the only predictor for Olympic winning times in several events. Now, although this predictor produces surprisingly good results, one would expect even better results if the male population of the world is replaced by predictors such as the population of participants or number of attempts as they are better indicators of how many times records were challenged. Considering the number of attempts as the major contributing factor to improvement of records and assuming that they occur according to a nonhomogeneous Poisson process, we can calculate the survival probability of the records.

Description of approach

Let $R > 0$ and $S > 0$ be two random variables with respective distribution functions $F_R(.)$ and $F_S(.)$. Suppose R, the record in a given sport, is subject an attempt S to break it. Then the record breaks if the value of S exceeds (or goes below) R. The value of S is a function of the type of the sport, number of participants, prize money, training, environmental factors such as temperature, rain, altitude, etc., and factors important to the athletes and the public. The value of R depends on factors such as the type, history, popularity of the sport, rewards or prizes, number of formal competitions, etc. The probability of breaking a record is, then,

$$P(S > R) = p = 1 - \int_0^\infty F_S(x) dF_R(x)$$

where p is the probability of breaking a record in a single attempt. Note that the calculations presented here can easily be modified for sports such as one-hundred-meter dash where breaking a record corresponds to $S < R$. When applying this model, one is frequently interested in the probability of breaking a record in a specified interval, say (0, t], where 0 represents the beginning of the period. Assuming that T is the length of time a record is (or will be) held, the probability of record being broken in the time interval (0, t], denoted by $F_T(t)$, can be obtained as

$$F_T(t) = P(T \leq t) = 1 - P(T > t) = 1 - L_T(t)$$

where $L_T(t)$ is the survival function defined as $L_T(t) = P(T > t) L_T(0)$

$= 1.$ If R is a record subject to a sequence of attempts $S_1, S_2, \ldots, S_n$, then $L_T(t)$ is given by

$$L_T(t) = \sum_{r=0}^{\infty} P(N(t) = r)\bar{P}(r)$$

where $\{N(t)\ t \geq 0\}$ is a general counting process of attempts occurring randomly in time and $\bar{P}(r) = P(\max (S_1, S_2, \ldots, S_r) < R)$, $r = 1, 2, \ldots, n$, with $\bar{P}(0) = 1$. If we further assume that the occurrence of the attempts is governed by a homogeneous Poisson process with rate λ, then we obtain

$$L_T(t) = \sum_{r=0}^{\infty} (e^{-\lambda t}(\lambda t)^r / r!)\ \bar{P}(r)$$

Note that, $\bar{P}(r)$ also represents the probability that a record survives the first r attempts. For independent and identically distributed, this reduces to

$$L_T(t) = \sum e^{-\lambda t}(\lambda t)^r / r!)(1-p)^r = \exp(-\lambda tp)$$

Thus, if the mean rate of attempts and the period of interest are given, $L_T(t)$ can be calculated for any given p.

Now, consider a more general situation in which

1. a record is assumed to be a random variable with distribution function $F_R(.)$;
2. the occurrences of attempts follow a stochastic point process P with a counting process $\{N(t), t \geq 0\}$, which develops over time and governs the values of attempts $\{S_n, n = 1,2,\ldots\}$; and
3. P and $\{S_n\}$ are independent.

Here, S_n denotes the nth independent and identically distributed random variable, realized at the moment N(t) first reaches n (S_n is the measured value of the nth attempt). Note that for this situation, breaking a record corresponds to the occurrence of the first record attempt (new high) in the S-sequence relative to the reference value R in the sense of the definition given below. In this approach, effects of other factors could be incorporated into P by increasing or decreasing the number of attempts or number of participants in a given period.

Definition: A first record with respect to R occurs at $t = t_{n_1}$ if

(a) $S_i < R$ for all $i < n_1$,
(b) $S_{n_1} > R$, in which case $T = t_{n_1} =$ the first record time, and $S_{n_1} = S_{(1)}$, the first record value. Alternatively,

$$T = \inf\{t : \max_{n \leq N(t)} S_n > R\}, \quad S_{(1)} = S_{N(T)}$$

Section 3 deals with calculation of survival probabilities assuming that the occurrences of attempts to break the record are governed by P, a Poisson process with time-dependent rate.

Record survival

Suppose that P is Poisson with time-dependent rate $\lambda(t) > 0$ and let

$$\Lambda(t) = \int_0^t \lambda(u)\,du$$

If $F_R(.)$ denotes the distribution function of R and $F_S(.)$ the distribution function of $\{S_n, n = 1,2,...\}$, then since $(T > t)$ if and only if $\max(S_1, S_2, ... S_n) < R$, the required probability $P(T > t)$ is given by

$$P(T > t) = \int_0^\infty \sum_{n=0}^\infty e^{-\Lambda(t)} \frac{(\Lambda(t))^n}{n!} (F_S(x))^n \, dF_R(x)$$

Note that $(F_S(.))^n$ is the distribution function of $\max(S_1, S_2, ..., S_n)$. This expression can also be written as

$$P(T > t) = \int_0^\infty \exp[-\Lambda(t)(1 - F_S(x)]dF_R(x)$$

If $R = R_0$ is given (e.g. R_0 is the present record), then

$$P(T > t \mid R = R_0) = \exp[-\Lambda(t)(1 - F_S(R_0))]$$

Now, it is clear that applications of these will require knowledge of both $F_R(.)$ and $F_S(.)$. However, there are two practically important cases discussed below where calculations can be carried out with less information and without numerical integration. Taking the viewpoint that the strength of a record in a given sport is measured by the number of attempts required to break it, these cases and the assumptions made seem reasonable.

Case A: Suppose that P has been observed throughout the time interval $(-\tau, 0]$, where 0 represents the present time. Suppose also that the largest value in this interval is used as a reference for determining further records. Then

$$F_R(x) = P(R < x) = \sum_{n=0}^\infty P[\max(S_1, S_2, ..., S_n) < x \mid N(\tau) = n]P(N(\tau) = n)$$

$$= \sum_{n=0}^\infty e^{-\Lambda(\tau)} \frac{(\Lambda(\tau))^n}{n!}(F_S(x))^n = \exp[-\Lambda(\tau)(1 - F_S(x))]$$

This yields

$$P(T > t) = \Lambda(\tau)[1 - \exp(-(\Lambda(\tau) + \Lambda(t)))] / [\Lambda(\tau) + \Lambda(t)]$$

With confidence given by the right-hand side of this expression, there will be no new maximum in (0, t] greater than the one in $(-\tau, 0]$. Thus, the survival probability depends only on the rate of attempts. This makes sense since breaking records depend not only on individual attempts but also on total number of attempts. The total number of attempts is related to the rate of population increase in general and increase in population of participants in particular. Note that since $D(t) = \Lambda(\tau) + \Lambda(t)$ is nondecreasing, for any $h > 0$

$$E(T) = \int_0^\infty P(T > t) = \int_0^h P(T > t) + \int_h^\infty P(T > t) = \int_0^h \frac{\Lambda(\tau)}{D(t)}[1 - \exp(-D(t))]dt + \int_h^\infty \frac{\Lambda(\tau)}{D(t)}[1 - \exp(-D(t))]dt$$

which implies that

$$\Lambda(\tau)[1 - \exp(-\Lambda(h))]\int_h^\infty \frac{dt}{D(t)} \le E(T) \le \Lambda(\tau) + \Delta(\tau)\int_h^\infty \frac{dt}{D(t)}.$$

Thus

$$E(T) < \infty \text{ if } \int_h^\infty \frac{dt}{D(t)} < \infty$$

It is interesting to observe that for a homogeneous Poisson process, the expected waiting time for a record to break is infinite. Recall that this is a well-known result in the theory of records for independent and identically distributed random variables.

Case B: This case is similar to case A except that, here, the exact number of attempts k that has occurred in the past is known (but the time span covered may not be known). Here, we are interested in breaking the mth best record $(m \le k)$. For example, in a certain competition, we may be interested in calculating the probability that a participant will break the third-best record. For this case, we have

$$F_R(x) = \sum_{j=k-m+1}^{k} \binom{k}{j} [F_S(x)]^j [(1 - F_S(x)]^{k-j}$$

$$dF_R(x) = m\binom{k}{m} [F_S(x)]^{k-m}[1 - F_S(x)]^{m-1}dF_S(x)$$

and

$$P(T > t) = \int_{-\infty}^{\infty} \exp[-\Lambda\,(t)(1 - F_S(x))]dF_R(x)$$

$$= m\binom{k}{m}\int_0^1 \exp[-\Lambda(t)u](1 - u)^{k-m}u^{m-1}du$$

For example, for m=1, we have $F_R(x) = (F_s(x))^k$ and therefore

$$P(T > t) = k\int_0^1 \exp[-\Lambda(t)u](1 - u)^{k-1}\,du$$

$$= k\,\frac{\exp[-\Lambda\,(t)]}{[\Lambda\,(t)]^k}\,\int_0^{\Lambda\,(t)} u^{k-1}e^u du$$

It is interesting to note that this case also covers the situation where the Lehmann alternative is satisfied by $F_R(.)$ and $F_s(.)$– that is,

$$F_R(x) = (F_S(x))^h \text{ or } 1 - F_R(x) = (1 - F_S(x))^h$$

for some h. Here, no reference is made to the past events, but R is assumed to behave as the $\max(S_1, S_2, ..., S_k)$.

Examples

Recall the following model used for the growth of the world's male population

Population in Year t = 1.6 exp [0.0088(t − 1900)]

This model can be approximated by a geometric increase with annual rate of exp(0.0088) = 1.0088388. Let us assume that the number of attempts is proportional to the population size at time t. Then using this expression, we have the following results:

a) The best record of a period of length of one hundred years has 80 percent chance of surviving an additional ten years.
b) The best record of a period of length of fifty years has 65 percent chance of surviving an additional ten years.
c) The best record of the period of length of ten years has 23.6 percent chance of surviving an additional ten years.

Suppose now that attempts occur randomly throughout $(-\tau, t]$. If n_1 attempts occur in the interval $(-\tau, 0]$ and n_2 attempts occur in the interval $(0, t]$, then the probability of no record in $(0, t]$ is

$$P(\max(S_1, S_2, \ldots S_{n_1}) = \max(S_1, S_2, \ldots S_{n_1 + n_2})) = \frac{n_1}{n_1 + n_2}$$

For this case corresponding to (a), (b), and (c) above, the ten-year survival probabilities are respectively $100/110 = 91\%$, $50/60 = 83\%$, and $10/20 = 50\%$, as no increase in participations or attempts is assumed. If, however, we assume a geometric increase of rate 1.0088388, then corresponding to (a), we have 86 percent. Note that the record of the last ten years may or may not be the same as the record of the last twenty years. This is one reason for the reduction in survival probability.

The example used above served as a demonstrating example. As mentioned earlier, a more realistic situation should consider a model for the population of participants or even the population of participants who have the potential to break records in a given sport. Good examples of this include participants who qualify for the Olympics games or participants who qualify for the Boston Marathon. The table and figure blow present data regarding the best times of the Boston Marathon together with the number of participants for the period 1970–2003. We chose 1970 as the starting year because this was the year during which a qualifying time was introduced. Using regression, the following relationship was found for the number of participants ($R^2 = 93.8\%$):

$$\text{Number of Participants in Year } t = -1294$$
$$+ 1088t - 57.5t^2 + 1.25t^3.$$

In this model, the data for the year 1996 was replaced by the average of the two neighboring times since this was the one-hundredth anniversary of the Boston Marathon and more than thirty-eight thousand runners were allowed to participate.

Using this model, the survival probabilities for the next five (2003–2008) and ten (2003–2013) years are respectively

$$P(T > 5) = 0.632 \quad \text{and} \quad P(T > 10) = 0.422$$

Moreover,

$$P(T > 10 | T > 5) = 0.667$$

As we know, the record set in year 1994 survived for twelve years and was broken in 2006.

We end this section by noting that data for the Boston Marathon during the period of thirty-four years includes six records. The maximum likelihood estimate of n, the number of independent and identically distributed attempts to produce six records, is 204. Also, the minimum value of n such that the expected number of records exceeds six for the first time is 227. From table 1, it is clear that the participation has increased steadily during the years. In fact, the total number of participants during the thirty-four years has been 287,539. If we take the number of participants for year 1970 (1,174) as the unit of participation, then we have $287,539/1,174 = 245$ unit of participants. If instead of 1970 we use the average number of participants for the period 1970–1974 as the unit, then we have $287,539/1,397 = 206$. Both cases provide good agreement with values of n mentioned above. If we add data for years 2004–2007, then we have thirty-eight years of data with seven records and $n = 565$.

Concluding remarks

In the analysis presented above, the number of attempts was used in a broad sense as the only factor contributing to the breaking of records. Clearly, there are many other factors that contribute to breaking of the records. However, as was pointed out earlier, one could account for these factors by changing the form of the $\Lambda(t)$ or by introducing additional terms.

Another interesting question relates to the idea of a possible ultimate record. In terms of what is discussed here, the ultimate record is the one that will survive forever (i.e., its survival probability is 1). Since every record will eventually be broken (provided that we can measure continuously, especially for increasing number of attempts), it is more practical to think of survival times such as fifty or one hundred years and a survival probability larger than, say, 90 percent, for a record to be considered an ultimate record. The approach is under investigation.

Data for Boston Marathon 1970–2003

Year	Winner	Time	Time (Minutes)	Number of Participants
1970	Ron Hill	2:10:30	130.50	1174
1971	Alvaro Mejia	2:18:45	138.75	1067
1972	Olavi Suomalainen	2:15:39	135.65	1219
1973	Jon Anderson	2:16:03	136.05	1574
1974	Neil Cusack	2:13:39	133.65	1951
1975	Bill Rodgers	2:09:55	129.92	2395
1976	Jack Fultz	2:20:19	140.32	2188
1977	Jerome Drayton	2:14:46	134.77	3040
1978	Bill Rodgers	2:10:13	130.22	4764
1979	Bill Rodgers	2:09:27	129.45	7927
1980	Bill Rodgers	2:12:11	132.18	5471

1981	Toshihiko Seko	2:09:26	129.43	6881
1982	Alberto Salazar	2:08:52	128.87	7647
1983	Greg Meyer	2:09:00	129.00	6674
1984	Geoff Smith	2:10:34	130.57	6924
1985	Geoff Smith	2:14:05	134.08	5595
1986	Rob de Castella	2:07:51	127.85	4904
1987	Toshihiko Seko	2:11:50	131.83	6399
1988	Ibrahim Hussein	2:08:43	128.72	6758
1989	Abebe Mekonnen	2:09:06	129.10	6458
1990	Gelinda Bordin	2:08:09	128.15	9412
1991	Ibrahim Hussein	2:11:06	131.10	8686
1992	Ibrahim Hussein	2:08:14	128.23	9629
1993	Cosmas Ndeti	2:09:33	129.55	8930
1994	Cosmas Ndeti	2:07:15	127.25	9059
1995	Cosmas Ndeti	2:09:22	129.37	9416
1996	Moses Tanui	2:09:15	129.25	38708
1997	Lameck Aguta	2:10:34	130.57	10471
1998	Moses Tanui	2:07:34	127.57	11499
1999	Joseph Chebet	2:09:52	129.87	12797
2000	Elijah Lagat	2:09:47	129.78	17813
2001	Lee Bong-Ju	2:09:43	129.72	15606
2002	Rodgers Rop	2:09:02	129.03	16936
2003	R. Cheruiyot	2:10:11	130.18	17567
Sum				**287539**

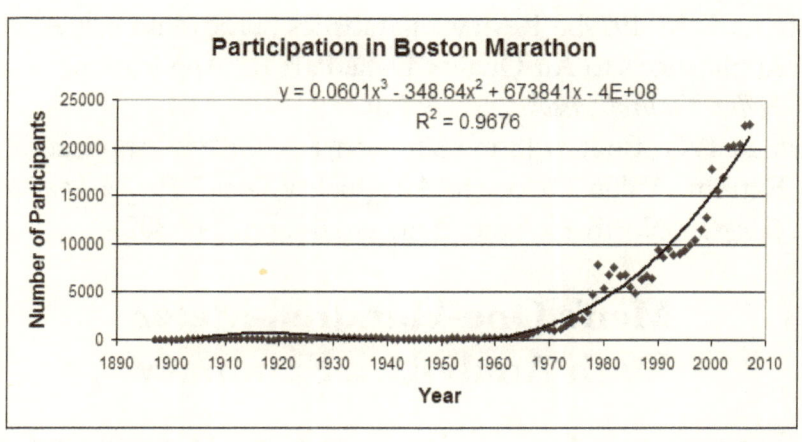

References

Cox, D. R., and Lewis, P. A. W. 1966. *The Statistical Analysis of Series of Events.* London: Chapman and Hall.

Davidson, A. 1984. "Modelling Excesses over High Thresholds, with an Application," in J. Tiago de Oliveria (ed.), *Statistical Extremes and Application,* D. Reidel, Dordrecht, pp. 461-482.

DuMouchel, W. 1983. "Estimating the Stable Index a in Order to Measure Tail Thickness." *Annals of Statistics,* 11, 1019-1036.

Gumbel, E. J. 1958, *Statistics of Extremes.* New York: Columbia University Press.

Hall, P. 1982, "On Estimating the Endpoint of a Distribution." *Annals of Statistics,* 10, 556-568.

Hill, B. M. 1975. "A Simple General Approach to Inference about the Tail of a Distribution." *Annals of Statistics,* 3, 1163-1174.

Noubary, Reza D. 2005. "A Procedure for Prediction of Sports Records." *Journal of Quantitative Analysis in Sports,* De Gruyter, vol. 1(1), pages 1-14, October.

Pickands, J. 1975. "Statistical Inference Using Extreme Order Statistics." *Annals of Statistics,* 3, 119-31.

Roberts, F. M. 1979a. "Review of Statistics of Extreme Values with Applications to Air Quality Data, Part I." Review, *J. Air. Pollut. Control. Assoc.* 29, 632-637.

Roberts, F. M. 1979b. "Review of Statistics of Extreme Values with Applications to Air Quality Data, Part II." Applications, *J. Air Pollut. Control. Assoc.* 29, 733-740.

Surman, P. G., Bodero, J., and Simpson, R. W. 1986. "Application of Extreme Value Theory to Air Quality Data." *Pacific Statistical Congress,* North-Holland, Amsterdam, pp. 299-302.

Men's One-Hundred-Meter Dash Analysis, a Summary

In recent years, because of the emergence of some exceptional athletes, prediction of athletic records has received a great deal of attention. For example, mathematicians have tried to model the improvements of records over time to forecast future records, including ultimate records. Records set in different sports shed light on human strengths and limitations and provide data for scientific investigations, training, and treatment programs.

Background

At the August 2009 world track and field competitions in Berlin, Usain Bolt, the Jamaican sprinting sensation, put on some amazing performances, shattering his own records in both the one hundred and two hundred meters, lowering both by 0.11 seconds to an amazing 9.58 seconds in the one hundred meters and 19.19 in the two hundred meters. The man is certainly on another level.

His time is the greatest improvement in the one hundred meters record since electronic timing began in 1968. Bolt is not done yet, and who knows how fast he can run? In fact, he thinks he can do better. He said this after winning the one hundred meters title at the same Olympic Stadium where Jesse Owens won four gold medals at the 1936 Berlin Olympic Games.

Several researchers have done studies to predict how fast a man can run one hundred meters. Most models are based on the idea that we are getting close to the limit because the amount of improvements is decreasing. Bolt's results changed this perception.

Bolt, who has set three records at the one-hundred-meter distance with times of 9.72, 9.69, and 9.58, is already looking to go far below some estimates. He thinks the world records will stop at 9.40. Some bookmakers are betting that Bolt will get there. The method for estimating the ultimate record presented below gives a 90 percent confidence interval for the ultimate record that has a lower bound of 9.40.

To measure the effect Bolt has had so far and may have in future, we calculated the lower bound excluding his three records. The result is 9.62, greater than the record he set, so Bolt has already beaten the estimated ultimate record based on other runners' times. We have also carried out some other probability calculations by excluding his records. They show how remarkable Bolt's performance has been. He has become the premier runner in the world and has changed our perception of human capabilities.

In the following sections, we will present methods we consider relevant for forecasting future records.

Methods based on trend analysis

Sports records have improved over the years, often faster than our expectation. To analyze the data, many investigators have used models made up of a deterministic component to account for the trend and a stochastic component to account for the random variation. Some attempts have also been made to estimate the limiting times (ultimate records) using models with exponential-decay deterministic component. Although useful, it is demonstrated that trend analysis would not produce meaningful performance estimates for future records.

Methods based on threshold theory

Consider the data for the men's one hundred meters after

January 1, 1977, when the International Association of Athletics Federations (IAAF) required fully automatic timing to the hundredth of a second. From January 1, 1977, to September 1, 2009, there are between thirty and forty data points, some being invalidated because of drug use and some being discarded (including Gay's 9.68 time on June 2008 during the 2008 U.S. Olympic Trials) because of wind speeds exceeding the IAAF legal limit. Here, the time between the two latest records set by Bolt and the time his recent record has stood to date is inordinately small. So rather than using an $m(n)$ that involves these values, here, we use $\sqrt{n}$ instead. This is a good choice as the rate of convergence is the same for both requirements. For the data considered, the closest integer to $\sqrt{n}$ is 6 ($m(n) = 6$). The top six legal times are

$$y_1 = 9.58, \ y_2 = 9.69, \ y_3 = 9.71, \ y_4 = 9.72, \ y_5 = 9.72, \ y_6 = 9.74.$$

set respectively by Bolt, Bolt, Gay, Powell, Bolt, and Powell. Using these for $y < y_7 = 9.77$, we get the tail model

$$F(y) = 6/33(9.77/y)^{-126.694}$$

Using this, $P(Y \leq 9.5)$ and $P(Y \leq 9.55)$ are respectively 0.0052 and 0.0102.

Bolt's effect

To measure the effect Bolt has had so far and may have in the future, let us calculate the same probabilities excluding his three records. The top six legal times are, then,

$$y_1 = 9.71, y_2 = 9.72, y_3 = 9.74, y_4 = 9.77, y_5 = 9.79, y_6 = 9.84.$$

set respectively by Gay, Powell, Powell, Powell, Greene, and Bailey. Using these $y < y_7 = 9.84$, we get the tail model

$$\bar{F}(y) = 6/33(9.84/y)^{-126.954}$$

The values of $P(X \leq 9.5)$ and $P(X \leq 9.55)$ are now respectively 0.00225 and 0.00433. These are significantly lower than the probabilities we get using Bolt's records. Also, we have $P(X \leq 9.58) = 0.0064$, which demonstrates that Bolt is in a different league.

Methods based on theory of records

The theory of records has not yet been fully exploited to address questions regarding the prediction of sports records because the results of the theory of records for independent and identically distributed sequences are not directly applicable to sports. In fact, in most sports, records occur more frequently than what the theory predicts. To account for this, one may treat the problem as if participation has increased with time or more attempts are made so that the probability of setting a new record was increased.

To predict the future records, I did develop a simple method using the following results of the theory of records for an independent and identically distributed sequence of observations (www.mathaware.org/mam/2010/essays/NoubaryRun.pdf) :

a. If there is an initial sequence of n_1 observations and a batch of n_2 future observations, then the probability that the additional batch contains a new record is $n_2/(n_1 + n_2)$.

b. For large n, $P_{r,n}$, the probability that a series of length n contains exactly r records is given by

$$P_{r,n} \sim \frac{1}{(r-1)!n}(\ln(n) + \gamma)^{r-1}$$

where $\gamma = 0.5772$ is Euler's constant. To illustrate the method, consider the one-hundred-meter data for the period 1912–2009. The number of records set is r = 20. The maximum likelihood method yields n = 317,884,920 attempts. This is found by maximizing $P_{r,n}$ with respect to n. It estimates the number of independent and identically distributed attempts required to produce twenty records.

Now, suppose, for example, that i is the geometric rate of increase per year because of increase in the number of attempts. This means that the number of attempts in any given year is assumed to be i times the number of attempts in the year before. The value of i can be found by solving the equation

$$1 + i + i^2 + i^3 + ... + i^{97} = 317884920,$$

where i^j represents the number of attempts in year j. We get i = 1.2013, which means 20.13 percent more attempts per year. To predict records for the future one and five years (at the time of that study year 2010 and the period 2010–2015), we replace $n_1 =$ 317884920 and $n_2 = (1.2013)^{98} = 63950619$ for the next year, and $n_1 = 317884920$ and $n_2 = (1.2013)^{98} + ... + (1.2013)^{102} = 477112800$ for the next five years in (a). This led to probability estimates of 0.1926 and 0.6001 for a new record in 2010 and the five years period 2010–2015, respectively.

Ultimate record

Prediction of ultimate records can be carried out using a model with an exponential-decay trend. It is also possible to consider simpler trends such as

$$y = (b + ct)/(1 + t) \text{ or } y = (a + bt + ct^2)/(1 + t + t^2)$$

and use the fact that as $t \to \infty$, $y \to c$. However, the application of such models usually results in predictions with large standard errors that are not useful or even acceptable. This section describes a method based on tail modeling using a certain number of exceptional performances. This avoids the above-mentioned problem and provides a confidence interval for the ultimate record based on the most recent performances or records.

Let Y and $Y_1, Y_2, \ldots, Y_n$ represent respectively a performance measure and a set of observed values for a given sport such as one-hundred-meter dash, where

$$Y_1 \leq Y_2 \leq \cdots \leq Y_n$$

Assuming that the distribution function $F(y)$ has a lower endpoint and certain conditions are satisfied, a level $(1 - p)$ confidence interval for the minimum of Y2 is

$$\{Y_1 - (Y_2 - Y_1)/[(1 - p)^{-k} - 1], Y_1\}.$$

It is shown that

$$\frac{\ln m(n)}{\ln[(y_{m(n)} - y_3)/(y_3 - y_2)]}$$

is a good estimate for k. For example, suppose that we wish to estimate the ultimate record for the one-hundred-meter dash. We consider the same data as in section 3.1. As in section 3.1, we need to choose an integer $m(n)$ depending on n such that $m(n) \to \infty$ and $m(n)/n \to 0$ as $n \to \infty$. Again, we choose $\sqrt{n}$ as it has the same rate of convergence for both requirements. For the data used, the closest integer to $\sqrt{n}$ is 6 ($m(n) = 6$) and the top six legal times are

$$y_1 = 9.58, \; y_2 = 9.69, \; y_3 = 9.71, \; y_4 = 9.72, \; y_5 = 9.72, \; y_6 = 9.74.$$

Using the above formula, we get $k = 4.419$, and a 90 percent one-sided prediction interval for the ultimate record is (9.40, 9.58). As mentioned earlier, Bolt believes that the world record will not go below 9.40, and the British bookmakers are betting that Bolt will get there.

Bolt's effect

Now, to measure the effect Bolt has had so far and may have in the future, let us calculate the prediction interval by excluding his three records. The top six legal times are, then,

$$y_1 = 9.71, y_2 = 9.72, y_3 = 9.74, y_4 = 9.77, y_5 = 9.79, y_6 = 9.84.$$

set respectively by Gay, Powell, Powell, Powell, Greene, and Bailey. Using the formula, we get $k = 1.113$, and a 90 percent one-sided prediction interval for the ultimate record is (9.62, 9.71). The lower bound is greater than the present record set by Bolt, so he has already exceeded what would have been predicted to be the limit of human performance. This shows how remarkable Bolt's performance has been and how this one runner can change our perception of human capabilities.

References

Brigstock-Barron, Rory. "Justin Gatlin Breaks Usain Bolt's 100m Record with 9.45 Second Dash on Japanese Television Show . . . But It Wouldn't Count." *Mail Online*, February 29, 2016. https://www.dailymail.co.uk/sport/othersports/article-3470075/Justin-Gatlin-breaks-Usain-Bolt-s-100m-record-9-45-second-dash-Japanese-television-wouldn-t-count.html.

International Association of Athletics Federations. "Usain Bolt Athlete Profile." https://www.iaaf.org/athletes/jamaica/usain-bolt-184599.

Noubary, Reza. "Tail Modeling and Prediction of Track and Field Records." *Journal of Applied Statistical Science* 16, no. 3 (2009): 287–292.

Noubary, Reza. "Tail Modeling, Track and Field Records, and Bolt's Effect." *Journal of Quantitative Analysis in Sports* 6, no. 3 (January 2010): Article 9.

Noubary, Reza. "What Is the Speed Limit for Men's 100 Meter Run?" In *Mathematics and Sports*, edited by Joseph A. Gallian, 287–294. Washington, DC: Mathematical Association of America, 2010. www.mathaware.org/mam/2010/essays/NoubaryRun.pdf.

Noubary, Reza, and Farzad Noubary. "Survival Analysis of the Men's 100 Meter Dash Record." *Applications and Applied Mathematics* 11, no. 1 (2016): 115–126.

Olympics Statistics. https://www.olympic.org/olympic-results.

CLOSING SUMMARY

- Traditionally, data values with high frequencies and their averages have been the focus of statistical modeling. This approach has rested on the important and powerful central limit theorem. Often, extremes, rare events, and records have been neglected–treated as outliers rather than as important information.

- The original Pythagorean Theorem of baseball was devised by Bill James in the 1980s. Several years later, Michael Jones and Linda Tappin of Montclair State University in New Jersey devised mathematically simpler alternatives to the theorem. Though useful, this rule has one exception that should never have occurred–DiMaggio's fifty-six-game hitting streak in 1941. Purcell calculated that to make it likely (probability greater than 50 percent) that a run of even fifty games will occur once in the history of baseball up to now (and fifty-six is a lot more than fifty in this kind of league), baseball's rosters would have to include either four lifetime 0.400 batters or fifty-two lifetime 0.350 batters over careers of one thousand games.

- Study of Larkey et al. (1989) reveals that different measures for hot hand lead to consider different players hot. Although the precise meaning of a term like *hot hand* is unclear, its common use implies a shooting record that departs from coin tossing with probability of success greater than that expected. To summarize the arguments against the hot hand, Tversky and Gilovich (1989), using several data sets, concluded that existing data does not support hot hand. They found that data for free throws provides no evidence that the outcome of the second shot depends on the outcome of the first one. Adams (1992), using data on eighty-three players, showed that the mean interval from making a field goal (n = 372) to making a field goal in nineteen NBA games did not differ from the mean interval from making to missing (n = 394), which further challenges assumptions regarding hot hand. As is pointed out in *The Skeptic's Dictionary*, the clustering illusion is the intuition that random events that occur in clusters are not really random events.

In an interesting paper, Wardrop (1995) presents many discussions concerning an inherent weakness in the methods used by Gilovich et al. (1985) and Tversky and Gilovich (1989a). Hooke (1989) discusses the inherent difficulty of using statistical methods to study complete phenomenon such as a game of basketball. Hale (1999) has discussed this issue and has raised several questions. He has argued that hot hand is an internal phenomenon and that the sense of being "hot" does not predict hits or misses. According to Hale (1999), there are three prominent arguments that conclude there are no hot hands in sports. In a more recent study, Koehler and Conley, in "The Hot Hand Myth in Professional Basketball," which appeared in the *Journal of Sport and Exercise Psychology*, have offered further evidence against hot hand in a unique setting–the NBA Long Distance Shootout contest. They have concluded that declarations of hotness in basketball are best viewed as historical commentary rather than as a prophecy about future performance. Having discussed opposing views, the final and perhaps more important question is, why do arguments for and against hot hand seem convincing?

In an attempt to answer this question, Wardrop (1995) has performed a very interesting analysis of data for Boston Celtics players. Wardrop used observed frequencies for pairs of free throws by Larry Bird and Rick Robey and the collapsed table (Wardrop 1995, table 1). In short, Wardrop has concluded that Simpson's paradox has occurred because the after-a-miss condition, when compared to the after-a-hit condition, has a disproportionately large share of its data originating from the far inferior shooter Robey. In other words, how you use the available data could lead to different conclusion.

- For a while, Bolt thought that he was not done yet and could run even faster. He said this after winning the one hundred meters title at the same Olympic Stadium where Jesse Owens won four gold medals at the 1936 Berlin Olympic Games. Records are breaking faster than what theory of records predicts because participation has increased with time, and as such, more attempts are made to set a new record.

To predict the future records, Noubary has developed a simple method using the results of the theory of records for an independent and identically distributed sequence of observations. However, the application of such models usually results in predictions with large standard errors that are not useful or even acceptable.

- Retired British major Walter Wingfield invented lawn tennis in 1873. Numerous aspects of tennis have been studied using both probability and statistical analysis, going back to Bernoulli's *Ars Conjectandi*. The British mathematician Ian Stewart wrote an amusing dialog along these lines called, "The Drunken Tennis-Player" for *Pour la Science*, the French version of *Scientific American*. Stewart's article was later reprinted in his book *Game, Set and Math*. Another nice discussion of mathematical modeling of tennis appears in *Mathematics and Sports* by Russian mathematicians L. E. Sadovskii and A. L. Sadovskii. Millionaire tennis player, tennis organizer, and tennis innovator James Van Allen promoted many of the alternative scoring systems.

Major Wingfield proposed that lawn tennis use badminton scoring–matches of fifteen points with players scoring only when serving. An alternate version of a tennis game was proposed by Van Allen, called a *no-ad game*. To avoid lengthy sets, James Van Allen promoted two variations on traditional scoring. Van Allen originally proposed a best-of- nine-point sudden-death tiebreaker. The United States Tennis Association used this version in the early '70s but later switched to what Van Allen disparagingly called a "lingering-death" tiebreaker–the first player to win seven points *and lead by two points* wins. A more radical variation on the rules for scoring sets is the Van Allen Streamlined Scoring System (VASSS), devised by Van Allen in 1958. To simplify the scoring and speed up matches, Van Allen scrapped games entirely. In fact, as is pointed out in an article titled "How Long Is an 11-Point Game?" djmarcusetasc.com (Wed. Feb. 10 10-39-01 1993) in most cases, the assumption of fixed probability has no effect on conclusion. Also, in a more detailed study titled "Does It Matter Who Serves First?" djmarcusetasc.com (Wed. Feb. 10 10-39-02 1993), which appeared in page 31 of Jan/Feb 91 *TT Topics*, the answer to the question posed in the title is stated as no based on the following argument.

Here, we make the common assumption that points are independent and identically distributed. Klaassen and Magnus (2001) analyzed 86,298 points from Wimbledon matches during the early 1990s and rejected the independent and identically distributed hypothesis, though they noted that the divergence is small enough (especially for strong players) that the hypothesis can be a reasonable approximation of what happens in practice.

However, as pointed out earlier it is not a serious one, and its effect on probability of winning or losing a game is not as pronounced as it is for tennis since the server has to hit her own side of the table first. In fact, as is pointed out in an article titled "How Long Is an 11-Point Game?" djmarcusetasc.com (Wed. Feb. 10 10-39-01 1993) in most cases, the assumption of fixed probability has no effect on conclusion. Also, in a more detailed study titled "Does It Matter Who Serves First?" djmarcusetasc.com (Wed. Feb. 10 10-39-02 1993), which appeared in page 31 of Jan/Feb 91 TT Topics, the answer to the question posed in the title is stated as no based on the following argument.

Closer to our investigations, the odds of winning a deuce game in tennis were analyzed using geometric series and later by solving a recurrence relation. The same method had been used by Ian Stewart in his 1991 book *Game, Set and Math*.

Here, we make the common assumption that points are independent and identically distributed. Klaassen and Magnus (2001) analyzed 86,298 points from Wimbledon matches during the early 1990s and rejected the independent and identically distributed hypothesis.

- Athletic performances and related issues have received considerable attention by physiologists, physical educators, and public. One aspect is concerned with the improvement of the records over time to address the question of forecasting the future records. In an attempt, Noubary (2005) considered the following simpler model instead that exhibits the same behavior as the logistic equation. Noubary (1994) has also suggested a trend analysis using a multiplicative model in place of an additive model. Berry (2002) proposed and used the male population of the world as the only predictor for Olympic winning times in several events.

- One study that appeared in *PLOS Medicine*, November 6, 2012, reports that people who engaged in physical activity had life expectancy gains of up to 4.5 years. Based on such studies, physical activities are recommended for anyone who can afford the time, energy, and the cost. Jason Dyck and his colleagues in University of Alberta in Canada found that red wine, nuts, and grapes have a complex called resveratrol, which improves heart, muscle, and bone functions; the same way they're improved when one goes to the gym. Details of this research can be found in the article by Natalie Roterman | Sep 15 2014, 04:51PM EDT.

Importance of Sports

The importance of sports in today's world is evident to everyone. Sports are one of the greatest unifying factors, and events such as the Olympics and the World Cup are indications of that. Different countries focus on different sports. To see this, consider the difference between sports in the two major sporting countries, the United States and Great Britain. The game that is most specifically associated with Great Britain is cricket. Although many other British games have been enthusiastically adopted by other cultures, cricket has found a home only inside the British Commonwealth. A fondness for cricket appears to mesh well with the British attitude. If so, however, the British may be losing some of their natural spirit, because a great majority of Brits have become much more enthusiastic about the eight months of the football (soccer) season than the four months of the cricket season. There are plenty of amateur football clubs, and professional clubs are a big business. Next to football, the chief spectator sports in England are horse racing and polo. Betting is common in all these sports.

The most popular sports in the United States are baseball, American football, and basketball. Not quite as popular but still widely watched is ice hockey. Baseball is an American pastime, and every year, millions of people attend games and watch it on television. American football is quite different from football in the rest of the world. British football, called soccer in America, is not yet as popular as the three main sports–football, baseball, and basketball.

Most American professional football games are played on Sundays, though recent years have seen some games move to Monday and Thursday nights, and college football games are played on Saturdays. All three sports have teams associated with major cities or states, and teams will sometimes move to another city or state that offers them more attractive terms. One hundred percent American in origin, basketball is popular at colleges and high schools, just like football and baseball. Statewide basketball tournaments are held annually. U.S. professional basketball teams are among the world's best, and the winner of the NBA championship is recognized as the champion of the world.

Obviously, there are many differences between British and American culture, even though the United States began as an English colony. A study of the divergence of cultures is also a study of how people with similar ancestry can develop different cultures even when separated for only a relatively short time. Such studies can provide insight into how cultures develop and provide insight into cultures different from our own.

In the United States, sports have gradually turned into entertainment, complete with cheerleaders, marching bands, and other associated activities. In both countries, where many people watch games on television, sports are a huge business as well.

In the United States education system, sports are sometimes considered more important than academics. Often a large percentage of a school's budget, in both high schools and postsecondary institutions, goes toward building sports facilities. Students on athletic teams and cheerleaders are often more popular than students who excel at academics. Many parents attend all of their children's sports events, though they may rarely or never request to observe their classes. This is not the case in Great Britain, where sports are more similar to religious and family traditions.

www.ingramcontent.com/pod-product-compliance
Lightning Source LLC
Chambersburg PA
CBHW021359210526
45463CB00001B/159